Introduction to Biotechnology
Volume 10 of
Basic Microbiology
a series edited by
J.F. Wilkinson

Introduction to Biotechnology

C.M. Brown DSc, FRSE
Professor of Microbiology

I. Campbell PhD
Senior Lecturer

F. G. Priest PhD
Senior Lecturer

All of Heriot Watt University,
Edinburgh

Blackwell Scientific Publications

OXFORD LONDON EDINBURGH

BOSTON PALO ALTO MELBOURNE

© 1987 by
Blackwell Scientific Publications
Editorial offices:
Osney Mead, Oxford OX2 0EL
(*Orders:* Tel. 0865 240201)
8 John Street, London WC1N 2ES
23 Ainslie Place, Edinburgh EH3 6AJ
52 Beacon Street, Boston
  Massachusetts 02108, USA
667 Lytton Avenue, Palo Alto
  California 94301, USA
107 Barry Street, Carlton
  Victoria 3053, Australia

First published 1987

Photoset by Enset (Photosetting),
Midsomer Norton, Bath, Avon
and printed and bound in the
United Kingdom by Billing & Sons
Ltd., Worcester

DISTRIBUTORS

USA and Canada
  Blackwell Scientific Publications
  Inc
  PO Box 50009, Palo Alto
  California 94303
  (*Orders:* Tel. (415) 965-4081)

Australia
  Blackwell Scientific Publications
  (Australia) Pty Ltd
  107 Barry Street,
  Carlton, Victoria 3053
  (*Orders:* Tel. (03) 347 0300)

British Library
Cataloguing in Publication Data

Brown, C.M.
  Introduction to biotechnology
  1. Biotechnology
  I. Title   II. Campbell, I.   III. Priest,
  Fergus G.
  660'.6   TP248.2

Library of Congress
Cataloging-in-Publication Data

Brown, C.M.
  Introduction to biotechnology
  (Basic microbiology; v. 10)
  Bibliography: p.
  Includes index.
  1. Biotechnology.   I. Campbell, I.
  II. Priest, F. G.
  III. Title.   IV. Series
  TP248.2.B76   1987   660'.6
  87–11641

ISBN 0–632–01139–4

Contents

Preface

This short textbook complements others in the *Basic Microbiology* series and assumes some background knowledge of biochemistry and microbiology. It was written with a student readership in mind but should prove equally useful to teachers in secondary and tertiary education. It draws on the experience of the authors, gained in teaching basic and industrial microbiology and biochemistry and other aspects of biotechnology to biologists and chemical engineers at Heriot–Watt University.

The treatment of subjects is necessarily selective but the book seeks to balance the traditional with the new biotechnologies and the science and engineering with their industrial applications and potential.

It is a pleasure to acknowledge the cheerful and efficient help given by Maggie Dunn in taming the word processors. The staff of Blackwell's Edinburgh office, especially Jane Starling and Nigel Palmer deserve a special mention for their patience and endurance. Finally, a warm thanks to Bob Campbell; without his enthusiasm and persistence this book would not have been written.

Chapter 1. Introduction

Biotechnology is one of the most used and abused words in modern biology with definitions to suit a wide range of views. This short book is concerned with biological systems (animal, microbial and plant) and their applications to agriculture, industry and the environment. Many applications are not new with examples in silage production, the traditional food industries and waste water treatment spanning several centuries. Current public awareness, however, stems from a number of relatively recent discoveries the most significant being the use of recombinant DNA technology to clone genes into suitable expression systems and the hybridization of spleen and myeloma cells to form hybridomas able to form monoclonal antibodies.

Between the traditional biotechnologies and the newer innovations lies a history of developments with a number of important milestones on the way. In any study of these developments it is clear that strategic and economic necessities as well as the inventiveness of biologists, chemists and engineers have given the necessary stimuli for advances.

Before 1800 the traditional fermentation industries were an art with brewing, baking, cheese making, etc. relying heavily on local skill and distribution of the products being confined to the area of production. There was little process control.

The nineteenth century saw some considerable advances with Pasteur's work on the 'diseases of beer' and the recognition of the involvement of yeasts in this most basic of fermentations. By 1896 Hansen was using pure cultures of yeast in the Carlsberg brewery in Copenhagen. Vinegar production was also widely practiced albeit without an understanding of the reasons why shallow trays and trickle filters were so successful in this process. The latter years of that century also saw the use of fermentation ethanol as a motor fuel and the microbial leaching of copper ores. In addition there were the well-known advances in medical microbiology with Koch's postulates and the recognition of the role of bacteria in human disease and the beginnings of an understanding of the immune system and the application of protective vaccines.

Developments in the early years of the present century include the first large scale anaerobic digestion plant for the treatment of sewage, the predecessor of modern biogas installations and the surface culture method of manufacturing citric acid for the food industry using *Aspergillus niger*. The World War of

1914–18 placed special demands on biotechnology and the Neuberg process for the production of glycerol (for nitroglycerine manufacture) with a 'steered' fermentation of *Saccharomyces cerevisiae* and the Weizmann process using *Clostridium* spp. for the production of solvents including acetone (for cordite manufacture) date from that period.

The next clear stimulus came some 20 years or so later with another world war. One of the most outstanding applications stemming from that time was the commercial production of penicillin, initially from surface cultures and later from submerged cultures of *Penicillium*. While being of great significance in its own right the production of penicillin on a large scale required the use of aseptic fermentation systems much larger than employed earlier and the development of modern fermentation technology dates from this time. In addition penicillin was the first antibiotic used therapeutically and revolutionized the treatment of infectious diseases. Shortly afterwards the pioneering work of Waksman resulted in the production of streptomycin. This work was of special significance since Waksman set out to screen micro-organisms isolated from the soil for antimicrobial properties. This successful strategy has been used by pharmaceutical companies ever since and has resulted in the isolation of many thousands of compounds, many from soil streptomycetes. Since the 1940s many other products of fermentation have been commercialized including the cephalosporin antibiotics, a number of amino acids (especially as a result of Japanese research), nucleotides, enzymes, vitamins and the gibberelins. In addition, the microbial transformation of plant steroids to form mammalian hormones has resulted in significant social and medical benefits. These developments were possible only with an increasing degree of complexity and sophistication of fermentation and extraction methods in order to produce this wide range of materials in an efficient and cost-effective way. Batch and fed batch fermentations have been the production methods of choice in these industries.

In the 1960s, concern over the availability of proteins for cattle and human consumption coincided with a plentiful supply of relatively cheap oil and related products. Thus single cell protein (SCP) was produced from hydrocarbons, methane and methanol and large scale continuous fermentations became popular. This presented the chemical engineers with new problems of mixing and aeration and the development of novel fermenters such as the pressure cycle and pressure jet systems. While variations in the price of oil and problems of acceptability of the products has had a profound influence on the commercial success of many of these processes, the basic engineering parameters have been established for large scale utilization of microbial cultures for future applications. Such systems are inevitably computer controlled.

In the last 25 years there have been many more applications of 'microbial biotechnology' including the production of high fructose syrups, polysaccharides such as xanthan gum and bacterial insecticides. The use of animal and plant tissues and cells in culture has a long history, for example the use of the former in the production of vaccines is extensive. Some of these applications are detailed in later chapters. In addition a large number of pharmaceuticals, enzymes and food colourings and flavourings are extracted directly from harvested plants. Slaughtered animals are also used as a source of pharmaceuticals such as insulin. Human blood is the source of a number of products including the clotting factor VIII and serum albumin, growth hormone is extracted conventionally from pituitary glands while the blood clot removing enzyme urokinase is extracted from urine.

To bring this brief history up to date, restriction and modification systems in bacteria were described in the early 1970s and the application of restriction endonucleases followed shortly afterwards while Milstein and Kohler published their now classical paper on hybridoma formation in 1975. New companies were developed to exploit these new discoveries and recent history has seen many academic entrepreneurs make and lose (often large) sums of money. These modern biotechnologies are usually heavily protected by patents. For example many of the basic steps in gene

Table 1.1. World markets for the main products of biotechnology.

Product	Sales ($ millions)
alcoholic beverages	34,500
cheese	21,000
antibiotics	6750
*fermentation alcohol	4000
diagnostic tests	3000
high fructose syrups	1200
amino acids	1125
bakers yeast	810
steroids	650
vitamins	495
citric acid	315
enzymes	300
vaccines	225
human serum albumin	185
insulin	150
urokinase	75
human factor VIII	60
human growth hormone	50
microbial insecticides	18

Figures for 1981 modified from Hacking (1986).
*Figure for 1984

Table 1.2. World sales of major products of biotechnology.

Product	Volume (1000 tonne)	Price ($ per tonne)	Value ($ millions)
Fuel/industrial ethanol			
USA	1000	576	576
Brazil	4500	576	2600
India	400	576	230
Others	400	576	230
Total			3636
High fructose syrups			
USA	3150	400	1260
Japan	600	400	240
Europe and others	200	400	80
Total			1580
Antibiotics			
Penicillin and synthetic penicillins			3000
Cephalosporins			2000
Tetracyclines			1500
Others			1500–2000
Total			8000–8500
Other products			
Citric acid	300	1600	480
Monosodium glutamate	220	2500	550
Yeast biomass	450	1000	450
Enzymes			400
Lysine	40	4000	160
Total			2040
Total of all products listed			15,256–15,756

Figures from 1983 from Hacking (1986).

cloning are covered by the Cohen–Boyer patent of 1980 and the 1981 patent awarded to Chakrabarty is significant in being the first to protect a 'man-made' micro-organism (even though in that case no gene cloning was involved). A final milestone worthy of note occurred in 1982 when the Eli Lilly company won safety approval from the Food and Drugs Administration of the USA for the use of 'human insulin' cloned into and produced by *Escherichia coli*. Many other products including the interferons, human and bovine growth hormones, and the hepatitis B virus surface antigen followed with many more to come.

In terms of annual world wide sales, the largest biotechnology industries by a large margin are those concerned with traditional fermentation products including the alcoholic beverages and cheese (Table 1.1) with antibiotics (total) and fermentation (fuel) ethanol each only about 10% of these. Biotechnology sales remain small compared with the chemical industry, which has products valued at \$700,000 million annually, and are also small compared with the size of the agriculture industry. For example one agricultural product, refined sugar (sucrose), had sales of approximately \$20,000 million in 1981. A more detailed breakdown of the world sales of some of the major products of biotechnology is given in Table 1.2. Fermentation ethanol used both as a fuel and as a chemical feedstock and high fructose syrup used as an alternative to inverted refined sugar are the only products competing directly with comparable materials from the (petro)chemical industries and agriculture. The others are either specialized microbial metabolites or are too complex and/or costly to be prepared by chemical synthesis. In general the successful products of biotechnology are either foods or food ingredients or are high value materials of use in human and animal health care where a high premium is possible. The new products using cloned genes are all likely to fall into this latter category in the foreseeable future.

Chapter 2. Microbial systems

Although biotechnology involves the potential use of all living forms, micro-organisms have played a major role in the development of this discipline and will continue to be used extensively for the foreseeable future. Reasons for this include the ease of mass cultivation of bacteria and moulds, the speed at which they grow, the cheapness of general medium constituents such as agricultural wastes, the massive diversity of metabolic types, which in turn gives rise to diversity of potential products, and the tremendous possibilities for genetic manipulation either in terms of strain improvement or new products.

Of the many thousands of microbial species (and there is probably an equal number yet to be described), relatively few are currently exploited. Some important products and producer micro-organisms are given in Table 2.1, and while this list is by no means exhaustive it emphasizes the scope of the technology. The principal micro-organisms involved are all chemo-organotrophs and derive their carbon and energy supply from the metabolism of organic compounds. Of the Gram-positive organisms, aerobic, endospore-forming bacteria of the genus *Bacillus*, some coryneform bacteria (*Corynebacterium* and *Brevibacterium*) and the filamentous bacteria (actinomycetes) particularly of the genus *Streptomyces* are well represented. Gram-negative types include acetic acid bacteria and xanthomonads. Various yeasts and filamentous fungi, are also involved. The relevant characteristics of these micro-organisms are that they should be nonpathogenic, easy to maintain and cheap to grow. *Escherichia coli*, *Bacillus subtilis* and *Saccharomyces cerevisiae* are being developed as hosts for genetically engineered genes and as such are potentially able to synthesize a range of products.

Isolation of micro-organisms

In this section, we shall assume that a market for a product has been identified and our task is to isolate a suitable producer-microorganism. The product must have the desired qualities with regard to activity, stability, nontoxicity, etc. but the nature of the producer-organism and product yield are not too important at this stage since they can probably be modified during strain development.

The demand for new biological products during the 1970s has

Table 2.1. Some important products from micro-organisms and their sources.

Product	Organism
Foods and beverages	
Bread	*Saccharomyces cerevisiae*
Beer and wine	*S. cerevisiae*
Blue cheeses	*Penicillium roqufortii*
Camembert and Brie cheeses	*Penicillium camembertii*
Soy sauce	*Aspergillus oryzae*
Sauerkraut and pickles	*Leuconostoc, Pediococcus* and *Lactobacillus*
Yoghurt and fermented milks	*Lactobacillus* and *Streptococcus*
Vinegar	*Acetobacter aceti*
Food additives	
Glutamic acid, lysine and amino acids	*Corynebacterium glutamicum* *Brevibacterium flavum*
Inosinic acid and ribonucleotides	*Corynebacterium glutamicum*
Vitamins	*Various yeasts and bacteria*
Enzymes—see Table 8.1	
Single cell proteins	
From alkanes	*Methylophilus methylotrophus* *Saccharomycodes lipolytica*
From milk whey	*Kluyveromyces fragilis*
From starch and other polysaccharides	*Fusarium graminearum*
From CO_2/sunlight	Various algae
Industrial chemicals	
Ethanol	*Saccharomyces cerevisiae, Zymomonas mobilis*
n-Butanol	*Clostridium species*
Citric acid	*Aspergillus niger*
Lactic acid	*Lactobacillus* species
Polysaccharides	
Xanthan	*Xanthomonas campestris*
Dextran	*Leuconostoc mesenteroides*
Alginates	*Azotobacter vinelandii*
Medical products	
Steroids	Mycobacteria and related bacteria
Hydroxylated steroids (cortisone, hydrocortisone, etc.)	*Rhizopus arrhizus, Curvularia lunata*
Antibiotics	
About 1000 compounds	Filamentous fungi
" 300 compounds	Actinomycetes
" 500 compounds	Nonfilamentous bacteria
Insect toxins	
Against lepidoptera larvae	*Bacillus thuringiensis*
Against mosquitos and blackfly	*B. thuringiens* var. *israeli* and *B. sphaericus*

encouraged more objective approaches to screening and isolating micro-organisms. Traditionally, large numbers of isolates were randomly screened in a search for the 'right' organism. The returns on this approach diminished fairly rapidly as it became apparent that the same organisms were being repeatedly assessed. For example the introduction of penicillin in 1941 was followed by an intensive search for β-lactam-type antibiotics and cephalosporin C was discovered in 1956. Until 1968 there were no further reports of novel β-lactam antibiotics and it was generally considered that few, if any remained to be found. However, since 1971, more than 12 natural, novel β-lactam antibiotics have been reported; the results of new, more objective isolation and screening programmes.

Principles of selective isolation

There are two aspects of selective isolation that should be considered. First, the production of some metabolites, such as antibiotics, amino acids or vitamins does not appear to confer a competitive advantage and therefore it is difficult to devise selective conditions for the isolation of such organisms in the laboratory. Instead, we must provide conditions that will permit the growth of unusual and novel micro-organisms and then screen them for useful metabolites. However, some products, for example enzymes, enable the producing organism to grow on a specific substrate and thus enrichment procedures can be adopted for selective isolation of potentially useful micro-organisms. Both aspects will be considered in the following section.

In attempts to isolate novel micro-organisms, microbiologists have sampled the most exotic habitats ranging from the ocean depths to deserts and from the smallest insects to the largest plants. But, if the growth media and conditions are kept essentially constant similar micro-organisms will be isolated whatever their origin. To isolate truly novel micro-organisms, we must not only sample unusual habitats but also use cultivation conditions that will repress the growth of common micro-organisms and encourage the rare ones. The four stages involved in the selective isolation of micro-organisms are presented briefly below.

1 *Selection of source material.* A variety of natural and manmade habitats should be sampled. In general, the more physically demanding the environment the more restricted the diversity of life. Highly alkaline or saline soils or thermal springs at 80°C will support fewer microbial taxa than garden soil but if thermostable or alkali-stable enzymes are required such habitats should be sampled. In general, as more exotic sites are sampled ranging from the Dry Valleys of Antarctica to the intestines of tadpoles or termites so new genera and species are discovered.

2 *Pretreatment of source material.* Selectivity can be introduced at this stage. Heat treatment which kills the predominant vegetative cell flora in soil is the classic procedure for selectively isolating spore-forming bacteria such as *Bacillus* spp. and actinomycetes. In the case of potential enzyme-producing organisms or those that degrade a particular molecule, such as a pesticide, either the original habitat can be baited with the substrate, thus enriching for the desired bacteria, or the material can be enriched in the laboratory. Repeated batch cultures with the substrate as sole carbon or nitrogen source will gradually increase the population of desirable strains. However the most powerful procedure is to use open-flow systems such as the chemostat and this is discussed later. After prolonged cultivation the best suited organisms or combination of organisms for the particular substrate will predominate, the others being 'washed out' of the system. Such enrichment techniques continue to provide many new and metabolically interesting isolates.

3 *Growth on laboratory media.* All laboratory media are to a degree selective, but the selectivity can be enhanced given a knowledge of the physiology of the required organism. Selective pressure in the form of temperature, pH or particular nutrient sources can be readily applied. In the case of the novel taxa there is no prior knowledge and the most effective policy is to suppress the growth of the common organisms in the hope that the rarer ones will then be able to grow. Antibiotics are a very useful aid in this context. Cycloheximide can be used to suppress moulds and yeasts and penicillin is more effective against a range of Gram-positive bacteria than Gram-negative. A combination of cycloheximide. mycostatin, chlortetracycline and methacycline has been used to selectively isolate *Nocardia asteroides* from soil. The length of incubation on these highly selective media is important. Colonies of *B. licheniformis* will appear in 18 h but for some actinomycetes incubation periods of one month to six weeks might be required.

4 *Novel screening procedures.* In the examples of antibiotics and pharmaceutical agents, novel screening procedures have been responsible for the isolation of bacteria that produce new antimicrobial compounds. For example, rather than use the standard bacteria such as *Staphylococcus aureus, Escherichia coli* and pseudomonads as indicators of antimicrobial activity, strains resistant or supersensitive to the common antibiotics have been used. In this way producers of the common β-lactam antibodies have been ignored and bacteria that produce new compounds isolated. Similarly, bacteria can be screened for induction of morphological aberrations or inhibition of differentiation rather than killing. Detection of pharmacological activities involves screens based on enzyme inhibitors, small organism (such as brine shrimps), animal tests and mouse behaviour.

Pure and mixed cultures

Microbiologists are taught to purify mixed cultures upon isolation so that experiments are conducted with monocultures. In nature, however, bacteria generally live in communities of varying complexity ranging from the dense populations found in animal intestines to barren soils in which microbes are few and limited in diversity. It is now being realized that for many biotechnological purposes, the obsessions with monoculture and single substrates may not necessarily be desirable and mixed cultures may have much to offer.

Many traditional fermentations, particularly food preparations, use mixed cultures, originally of variable composition but increasingly standardized inocula are used. A prime example of the benefits of a mixed culture are the fine wines of the world produced from the natural flora of the grapes which will probably never be equalled by the use of pure yeast cultures, although the latter can provide uniform wines of reasonable quality. It should be stressed that the micro-organisms may not all be present at the same time, frequently a succession is involved as the environment changes due to the excretion of metabolites. Thus in the preparation of Belgian lambic beer, malt wort is allowed to ferment spontaneously and is colonized by various Enterobacteriaceae. After about 40 days, the enteric bacteria are replaced by *Saccharomyces* yeasts. A third period of fermentation then starts characterized by pediococci and then 6 to 8 months from the start of the fermentation *Brettanomyces* yeasts comprise the fourth fermentation period.

Mixed cultures are being exploited in aspects of modern biotechnology because they offer culture stability and resistance to contamination, higher biomass yields than monocultures, and the ability to metabolize toxic waste materials that are often recalcitrant to single strains. The higher biomass yields are particularly relevant to single cell protein production where mixed bacterial cultures growing with methane as carbon source have been shown to provide higher yields than pure cultures. Quality control could be a problem however if the fermentation is not carefully controlled. Microbial degradation of organic compounds is frequently based on consortia. This can be readily demonstrated by inoculating a chemostat with soil and providing an aromatic hydrocarbon as carbon source. After a short while a stable community of micro-organisms is usually established. A typical case might be the two organisms shown in Fig. 2.1; a *Rhodococcus* ('*Mycobacterium*') *rhodochrous* strain that grows at the expense of acetyl fragments generated by the β-oxidation of the side chain of dodecylcyclohexane, and an *Arthrobacter* strain able to utilize the resultant cyclohexane acetic acid. In many such examples several micro-

organisms may be involved that alone would fail to grow on the
supplied substrate. The importance of these consortia for the com-
posting of plant material and the biodegradation of waste materials
cannot be over emphasized.

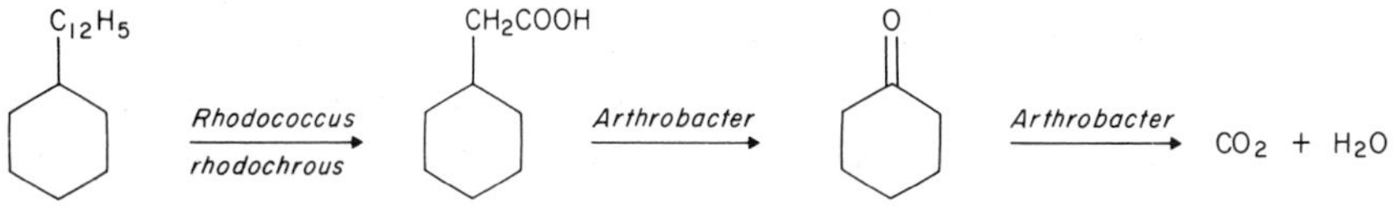

Fig. 2.1. The complete degradation of dodecylcyclohexane by the synergistic
attack of two bacteria (adapted from Feinberg *et al.* 1980).

The improvement of industrial micro-organisms

Having identified an organism that makes a marketable product
or is involved in a new process the cost of the product is critical.
If it is expensive compared with existing products and processes,
unless it has considerable advantages it will not be used. The key
feature therefore is to reduce production costs by strain improvement either to increase yield or to reduce the cost of manufacture
by offering, for example, enhanced growth on cheaper substrates
or simpler purification processes. Thus strain modification is
crucial to modern biotechnology. Before we can consider the
various approaches to strain improvement it is important to understand the principles of the regulation of protein synthesis and
enzyme activity. Protein synthesis occurs in three stages; transcription, translation and post-translational modification all of which
may be controlled within the cell. Applications of gene cloning
in microbial systems are discussed in Chapter 3.

Transcription and its control

Transcription describes the binding of RNA polymerase to a
specific site on the chromosome (promoter) and the synthesis of
a messenger (m) RNA complementary to a strand of the DNA. It
is at this stage that the primary control of protein synthesis is
effected in prokaryotes and lower eukaryotes, although evidence
for translational control is accumulating. Although transcription
is a universal process, the molecular details vary considerably
even among bacteria. In *E. coli*, RNA polymerase occurs in essentially a single form comprising a core enzyme of four polypeptides
responsible for mRNA synthesis and two additional polypeptides,
the sigma and rho factors that ensure transcription initiates and
terminates at the correct sites. In *Bacillus subtilis* and probably
streptomycetes on the other hand, RNA polymerase is more
heterogeneous and the core enzyme can combine with a range of

sigma factors each of which confers specificity for initiation at particular promoters. These factors are principally associated with sporulation and in *B. subtilis* there exists a 'cascade' of sigma factors that associate with RNA polymerase and regulate expression of sets of genes temporarily during differentiation. During exponential growth, RNA polymerase combines with a sigma factor of 43,000 molecular mass (σ^{43}, previously σ^{55}) and vegetative genes are selectively transcribed. As the culture enters stationary phase and during sporulation, sigma factors of 37,000, 29,000 and 28,000 have been identified and are responsible for the transcription of sporulation-associated products during differentiation. The situation in *Streptomyces* is unclear but factors resembling σ^{43} and σ^{37} have been identified.

Since the rate of chain elongation is essentially constant at constant temperature, transcription can be regulated at just three points; RNA polymerase binding, progress from promoter into the gene and termination.

RNA polymerase binding requires specific base sequences within the promoter at loci situated about 10 base pairs (bp) upstream from the initiation of transcription (−10 or Pribnow box) and 35 bp upstream (− 35). In *E. coli*, the canonical sequences for RNA polymerase binding are shown in Fig. 2.2 and the nearer

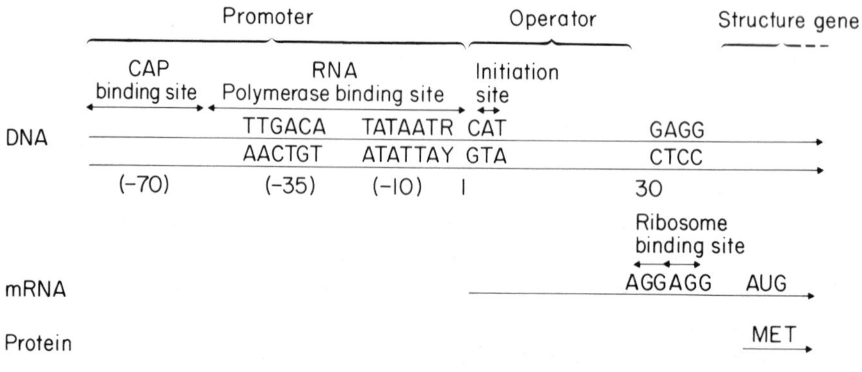

Fig. 2.2. Regulatory regions in protein synthesis in *Escherichia coli*. RNA polymerase binds to the promoter region of the DNA, the strength of the binding being dependent on how closely the −35 and −10 sequences resemble the optimal shown here. Promoters that come under catabolite repression control have an additional binding site for catabolite activator protein (CAP). In the presence of cyclic AMP, CAP interacts with the DNA and enhances RNA polymerase binding to the promoter. Inducible and repressible operons contain an operator sequence between the promoter and structural gene, the site of recognition by an activator (positive control) or repressor (negative control) protein (*see* text). The mRNA contains a sequence complimentary to the 3''-end of the 16S rRNA (The Shine–Dalgarno sequence) some 10 bases upstream from the translation initiation site, AUG. (Regulatory sequences differ in *Bacillus subtilis* and *Streptomyces*.)

the actual sequences are to these optima the stronger the promoter. The equivalent sequences for the σ^{43} promoters in *B. subtilis* are TTGACA (-35) and TATAAT (-10). This close resemblance to the *E. coli* sequences at least partly explains the expression of *B. subtilis* genes in *E. coli* and the (less efficient) transcription of *E. coli* genes in *B. subtilis*.

As promoter sites depart from the canonical sequences so RNA polymerase binding becomes less efficient and the promoter 'weaker'. Thus the -10 sequence of the lactose (*lac*) operon promoter, TATGTTG differs from the canonical sequence TATAATR at two positions and can only be used by RNA polymerase when cyclic 3,5-adenosine monophosphate (cAMP) together with the activator protein CAP are present in the cell. This is part of the catabolite repression control system in *E. coli*. Catabolite repression refers to the inhibition of expression of operons for the catabolism of poor (i.e. low-growth rate sustaining) carbon sources by glucose or some other rapidly metabolized carbon source. This coordination of carbon metabolism avoids wasteful synthesis of a variety of catabolic enzymes when the organism is presented with several carbon sources at once and ensures that the most readily metabolized source is used first. In slowly growing *E. coli*, the intracellular concentration of cAMP is high and together with CAP enhances RNA polymerase binding at catabolite sensitive operons such as *lac* if the operon is induced. Rapid growth provides a low cAMP concentration and such operons cannot be transcribed even in the presence of inducer, thus preventing the wasteful synthesis of these catabolic enzymes. Similarly, the addition of glucose to a culture growing slowly on lactose leads to a rapid decline in cAMP and loss of catabolite sensitive operon expression.

The cAMP/CAP complex enhances the affinity of RNA polymerase for the promoter by interacting with an adjacent upstream site and destabilizing the duplex, thereby increasing the efficiency of open promoter formation. Indeed *lac* promoter mutations that generate the canonical sequence at -10 are no longer sensitive to catabolite repression. It should be emphasized that although catabolite repression is a common phenomenon, the molecular details vary and cAMP is not involved in the mechanics of catabolite repression in bacilli nor probably in streptomycetes.

Transcription can also be controlled by placing a regulatory protein at a site (operator) between the promoter and structural gene(s) (Fig. 2.2). An operator that allows free passage of RNA polymerase but is progressively closed by a protein repressor defines negative control and positive control refers to an operator that prevents RNA polymerase from transcribing the gene but is opened by binding an activator (Fig. 2.3). Inducible, catabolic operons of *E. coli* in which the substrate or an intermediate of the

No expression   Expression

(a)   Inducible operon, negative control

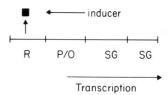

                                                     Transcription

(b)   Inducible operon, positive control

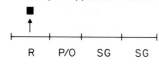

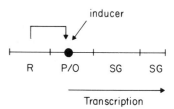

                                                     Transcription

(c)   Repressible operon, negative control

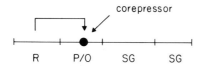

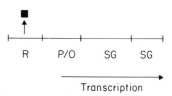

                                                     Transcription

(d)   Repressible operon, positive control

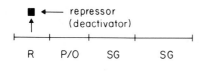

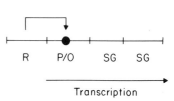

                                                     Transcription

Fig. 2.3. Control of gene expression. A regulatory gene *R* directs the synthesis of a protein that interacts at the operator site *O* adjacent to the promoter *P* and regulates transcription of the structural genes, *SG*.
(a) Inducible operon under negative control (for example *lac*). *R* codes for a repressor that inhibits transcription. In the presence of inducer, the conformation of the repressor is altered such that the affinity for *O* is much reduced and transcription proceeds.
(b) Inducible operon under positive control. *R* codes for an activator that can only enhance transcription in the presence of inducer.
(c) Repressible operon under negative control. The repressor (aprorepressor) is converted to an effective conformation by interaction with the product of the operon (corepressor) and prevents transcription.
(d) Repressible operon under positive control. The activator of transcription is only effective in the absence of the product of the operon (repressor). (Redrawn from Priest 1984.)

pathway induces expression of the relevant operon include the negatively controlled *lac* and histidine utilization systems while the arabinose and maltose operons are positively regulated. All biosynthetic (repressible) operons in which the product of the pathway represses synthesis of the relevant enzymes studied to date are negatively controlled.

A third mode of transcriptional control operates in amino acid biosynthetic operons of *E. coli* and other bacteria. Termed attenuation, it involves the premature termination of transcription at a site between the operator and structural genes giving rise to a truncated mRNA. Attenuation occurs when the relevant amino acid is present in the environment at insufficient levels to effect repression of the operon and offers fine control of protein synthesis.

Translation and its control

In prokaryotes transcription and translation are usually coupled and as the message is synthesized so it is translated into protein. An exception seems to be extracellular proteins which are exported across the membrane as they are synthesized and the mRNA/ribosome complex must be transported from the chromosome to the cytoplasmic membrane before translation can proceed.

The rate of peptide chain formation is probably invariant under standard growth conditions and therefore regulation of translation must focus on the initiation step and stability of the mRNA, although other factors such as codon usage by the translation machinery, availability of substrates and possible toxicity of product will play a part. Translation does not appear to be a major control point in protein synthesis in prokaryotes but is probably more important in eukaryotes.

Initiation requires a specific interaction between the 16S ribosomal (r)RNA of the small ribosomal subunit and the ribosome binding site (RBS) of the message. A consensus sequence, AGGAGGU (termed the Shine–Dalgarno [SD] sequence after its discoverers) centred about 10 nucleotide residues from the initiation codon (AUG) is complementary to the 3'-terminal sequence of the 16S rRNA and necessary for formation of the initiation complex together with N-formyl methionyl-tRNA$_f^{MET}$ and the initiation factors IF-1, IF-2 and IF-3. Although regulation of initiation of translation is a controlling parameter for the synthesis of ribosomal proteins in *E. coli,* the principal interest of the biotechnologist is to render the initiation event as efficient as possible. The RBS of different genes may vary by nearly a thousand-fold in relative efficiency and it is important that the translation initiation signal should be modified to provide (1) an efficient SD sequence for the organism in question and (2), the optimum space between the SD and the initiation codon (Fig. 2.2).

Gram-positive cells are not capable of translating mRNAs from Gram-negative bacteria efficiently but Gram-negative bacteria can generally translate Gram-positive messages. Hence *E. coli* can express *B. subtilis* genes but not vice versa. It seems that Gram-positive bacteria require greater complementarity between the SD sequence and 3'-end of the 16S RNA than is necessary in Gram-negative bacteria. Thus the conserved Gram-positive sequences will operate in *E. coli* but the more stringent translational apparatus of *B. subtilis* will not initiate on most *E. coli* mRNAs in which the SD sequence may deviate substantially from the consensus sequence.

Most mRNAs in bacteria are unstable with half-lives of 2–4 minutes at 37°C. Improvement in productivity from a gene could be gained by stabilizing the message.

Aspects of protein synthesis in eukaryotes

Since biotechnology encompasses the cloning and expression of eukaryotic genes in bacteria it is important to describe briefly how protein synthesis in eukaryotic cells differ from that in pro-karyotes. The primary transcript from the eukaryotic chromosome is not mRNA but a high molecular weight 'heterogeneous nuclear' (hn) RNA. This material is processed by the removal of segments (introns) and the 'splicing' of the remaining sequences (exons) to

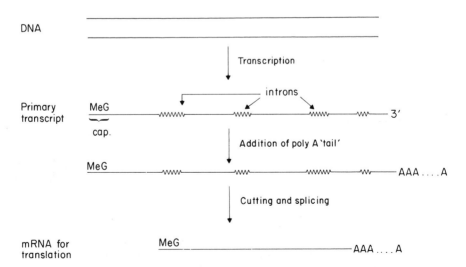

Fig. 2.4. Processing of messenger RNA in eukaryotic cells. Before the primary transcript is released from the DNA its 5'-end is modified by the addition of a 7-methyl guanosine 'cap' (MeG) and subsequently a stretch of adenine residues is added to the 3'-end. The intervening sequences are then excised and the final transcript is produced.

provide the message. Further modification in the form of the addition of a methylated oligonucleotide 'cap' to the 5'-end and a stretch of adenine residues (poly[A]) to the 3'-end complete the processing (Fig. 2.4). Thus eukaryotic genes complete with introns are not expressed in bacteria nor often in lower eukaryotes such as yeast because the enzymes necessary for processing are absent.

Protein secretion

An important aspect of protein synthesis is the secretion of proteins across the cytoplasmic membrane of bacteria or the membrane of the endoplasmic reticulum of eukaryotes. This is particularly important for biotechnologists since the commercial advantages of a secreted protein product include higher yields and ease of recovery and purification compared with cytoplasmic proteins. The process of secretion has been remarkably conserved throughout evolution and is similar in both prokaryotic and eukaryotic cells (Fig. 2.5). In brief, secreted proteins are synthesized with amino-terminal peptide extensions of 20–30 amino acids, the signal or leader peptide. As it emerges from the large ribosomal subunit, it binds the signal recognition particle (SRP) which temporarily halts peptide chain elongation. The mRNA/

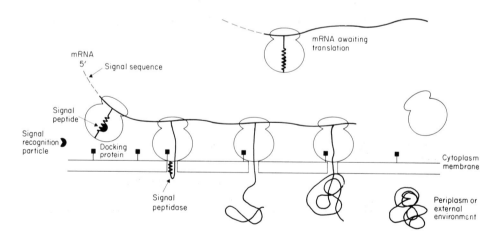

Fig. 2.5. Schematic diagram of protein secretion. The overall scheme is similar in eukaryotic and prokaryotic cells but the fine details may vary. As the mRNA is transcribed, ribosomes bind to the 5'-end but translation is halted after a short while by the signal recognition particle. Interaction of the ribosome with the docking protein on the inner surface of the membrane allows translation to proceed. The signal peptide binds to the inner surface of the membrane and the hydrophobic region of the peptide is inserted into the membrane. Signal peptidase removes the signal peptide and the nascent protein is transported across the membrane as it is synthesized. On completion of translation the ribosome is released from the membrane and the protein assumes its correct conformation. (For further details see Priest 1984, Randall and Hardy 1984.)

ribosome/SRP complex migrates to the inner side of the membrane where it is bound and interacts with the 'docking' protein which triggers resumption of translation. As the peptide chain elongates it is transported across the membrane and is released on the outer side where it assumes its native configuration. 'Stop' sequences in the protein halt translation and can give rise to integral membrane proteins that span the membrane.

The process is very similar in bacteria although the SRP and docking proteins have yet to be unambiguously identified. Secreted proteins in Gram-positive bacteria generally diffuse through the cell wall to become totally extracellular but in Gram-negative bacteria they usually accumulate in the periplasm between the cytoplasmic and outer membranes. It follows, in theory at least, that virtually any protein could be made in secreted form by supplying it with the correct signal peptide for the host organism. This has not been fully realized and although strains of *B. subtilis* that secrete interferon and insulin have been constructed, success has only been achieved with naturally secreted proteins. It may be that the structure of cytoplasmic proteins is not amenable to secretion for some reason.

Regulation of enzyme activity

Cells control their metabolic activity according to environmental stimuli by inducing and repressing the synthesis of proteins. A second level of control is the direct modulation of enzyme activity. In bacteria this commonly occurs in biosynthetic pathways particularly those associated with amino acid synthesis in which the product inhibits the activity of an enzyme early in that particular pathway by feedback inhibition. In the case of branched pathways leading from a single precursor to two or more products the systems become more complicated. The lysine, methionine, threonine pathway illustrates the various strategies (Fig. 2.6). It is usual for the product of a branch of the pathway to inhibit the first enzyme of the branch. Thus lysine inhibits step 8, threonine step 6, and methionine step 4 in Fig. 2.6. However, this could lead to the unwanted accumulation of an intermediate, in this case homoserine, so the intermediate, and the products of the pathway generally inhibit the very first step of the scheme either singly or cumulatively. Obviously to increase and direct metabolic flux through such pathways, these restrictions must be removed where necessary.

Strain improvement

The aim of strain improvement programmes is to destroy the regulatory mechanisms of an organism such that maximum

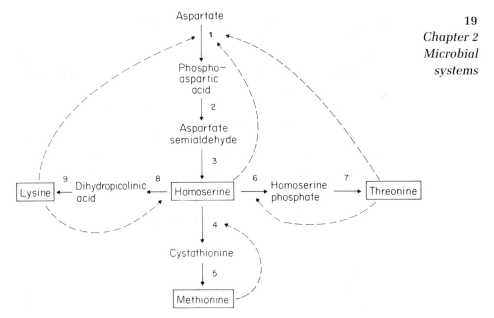

Fig. 2.6. Control of the pathway leading to the synthesis of lysine, threonine and methionine. Dashed arrows indicate feedback inhibition. For details see text.

metabolic energy is devoted to a single product, either a particular protein or the product from a metabolic pathway. There are two approaches to strain improvement. The traditional mutation and selection/screening programmes involve random mutagenesis of the producer organism followed by screening survivors for enhanced yield. These organisms are then further improved in successive rounds of mutagenesis and screening. This approach has the advantages that it is generally relatively rapid and has a good track record. In the penicillin programme for example, it was particularly effective. Similarly, α-amylase yields from *Bacillus subtilis* have been improved thousand-fold by mutation and screening. On the other hand, techniques involving gene cloning have resulted in *E. coli* strains in which some 20% of the protein-synthesizing capacity of the organism is devoted to a single product. The choice of which avenue to follow will largely depend on the nature of the producer-organism and the product. Four general categories can be identified but most commercial examples would probably overlap the boundaries.

(1) Product of a single gene or operon; organism simple to grow. A typical example could be an enzyme from a bacterium such as *B. subtilis*. In the past this would have been improved by mutation and screening but with new developments in gene-cloning it is likely that a recombinant-DNA approach would be adopted today. (2) Product of a single gene or operon; organism pathogenic or difficult to grow. Such a category could include plant or animal products and the route to efficient manufacture

would be through cloning the gene(s) into a suitable host cell. (3) Product of several, unlinked genes; organism easy to grow. The commonest example in this category might be antibiotic synthesis. Numerous genes which may be scattered around the chromosome (although this now seems unlikely) are often responsible for such secondary metabolites which makes gene cloning impractical and favours the traditional mutation and selection approach. Only when the producer organism is of outstanding commercial or scientific importance, as in the case of *Streptomyces* is genetic analysis worth considering. (4) Product of several unlinked genes: organism pathogenic or difficult to grow. Such a product is probably best shelved until a more suitable source is discovered.

Mutation programmes

Traditional mutation and selection programmes involve blindly mutating a producer organism in the hope that regulatory genes will be sufficiently damaged to remove inducer requirements, catabolite repression, feedback inhibition and other constraints on protein synthesis and activity. At each stage, screening reveals strains with increased productivity and these are further mutated, often using different mutagens, in an attempt to modify different loci and further increase yield. If the organism has a system of genetic exchange, high yielding mutants of different lineages are recombined to generate further yield improvements and as we learn more about protein synthesis so the mutation and screening stages can be directed at the removal of specific restrictions. Two successful examples will suffice to outline the approaches that have been used.

Serratia marcescens has been modified for the production of histidine by mutation. Nine structural genes code for enzymes that result in the synthesis of histidine from phosphoribosyl-pyrophosphate and ATP (Fig. 2.7) Moreover, exogenous histidine can be used as a source of carbon and nitrogen through the catabolic pathway involving histidase and urocanase. Wild-type strains regulate histidine biosynthesis by feedback repression of

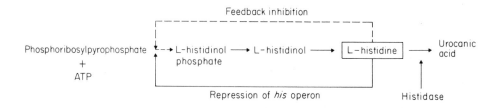

Fig. 2.7. Pathway of histidine biosynthesis in *Serratia marcescens*. Adapted from Enei and Hirose (1984), with permission.

the first enzyme of the pathway and end product repression of transcription of the complete operon. A histidase deficient mutant (Hd-16) was isolated first, as unable to grow on histidine as a sole source of nitrogen. This prevented degradation of over-produced histidine. Feedback repression can often be removed by isolating mutants resistant to analogues of the amino acid. These analogues bind to the regulatory site in the enzyme which render it nonfunctional and the analogue is thus toxic. By selecting for resistance to 2-methyl histidine, a mutant (MHr131) no longer susceptible to feedback repression was isolated. Similar mutants, resistant to ethionine, β-2 thienylalanine and 6-mercapto guanine have lost feedback repression of methionine, phenylalanine and inosinic acid biosynthesis respectively in various bacteria. Mutations causing constitutive synthesis of the *his* operon were obtained by mutating and selecting for growth on minimal medium with carbon source and 1,2,4-triazole-3-alanine (TRA). This molecule, like histidine represses expression of the *his* operon and thus the organism cannot grow, but constitutive mutants (for example Tr 142) will grow normally. Other ploys for isolating constitutive mutants include for example growth on substrates for an enzyme that do not induce its synthesis such as phenyl-β-D-galactoside for β-galactosidase synthesis in *E. coli*. Individually, the two mutations did not substantially increase histidine biosynthesis, but when the TRA-resistance was transduced into the MH-resistant strain and a double mutant (H2892) constructed, histidine production was 20 g/l (*see* Table 2.2). The strain grew slowly however and was unstable probably due to low intracellular ATP concentration resulting from the ATP being channelled into histidine biosynthesis. A derivative resistant to 6-methylpurine was isolated which overproduced ATP and showed more rapid growth and was stable during histidine production.

Table 2.2.

Strain	Repression	Inhibition	Histidine	Generation time (h)	L-histidine (mg/h)
9000	+	+	+	1.0	0
Hd-16	+	+	−	1.0	0
MHr131	+	−	−	1.2	1.0
Tr142	−	+	−	1.3	1.5
H2892*	−	−	−	2.6	20.0
MPr-90	−	−	−	1.9	22

*H2892 was constructed from a transductional cross between HD-16 and MHr131.

The second example of successful mutation and selection programmes involves production of the extracellular amylase from *B. subtilis*. This enzyme is constitutive but catabolite repressible and high yielding mutants can be obtained by screening for zones of starch hydrolysis around colonies. There is no direct selection for mutants that overproduce amylase, consequently, one strategy is to isolate mutants that are resistant to antibiotics that interfere with transcription, translation and secretion in the hope that they will be pleiotropically altered in extracellular enzyme production. Thus rifampicin and streptolydigin resistant mutants have altered RNA polymerases, ribosomal inhibitors can be used to obtain mutants with altered large and small subunits and antibiotics and detergents that affect the cell wall and membrane can be used to obtain mutants with deranged cell envelopes. Yamane and Maruo (1980) adopted this type of approach and isolated a variety of individual mutants with slightly increased amylase yield. However, when these individual mutations were assembled in a single strain by DNA-mediated transformation, a 1500-fold increase in productivity was realized. In other bacteria conjugation or protoplast fusion can be used to effect the transfer of alleles into a single host strain.

Chapter 3. The principles of gene cloning

Gene cloning in bacteria

The advent of gene cloning techniques in micro-organisms raised exciting possibilities for the geneticist interested in increasing product yields. It became possible to analyse gene structure in detail, use *in vitro* mutagenesis to mutate at specific sites, splice genes onto strong promoters and ribosome binding sites and to place genes in multi-copy plasmids to enhance productivity through gene dosage. Aspects of genetic engineering will be covered briefly here, for a more thorough treatment the reader should consult specialist reviews and texts; in particular Old and Primrose (1985) and Glover (1984) are recommended.

The strategies of gene cloning are outlined in Fig. 3.1.

There are essentially four stages.

1 *Preparation of the gene.* Cloning bacterial genes is generally achieved from total chromosomal DNA preparations ('shotgun' cloning) by cleaving the DNA with a restriction endonuclease that generates fragments of approximately 4 kilobase pairs (kb) each with a 'sticky' complementary single stranded end. Eukaryotic genes contain introns that are not processed in bacteria so DNA for cloning is generally obtained as a reverse transcriptase generated copy (c) DNA of the relevant mRNA (Fig. 3.1). DNA may also be synthesized synthetically if the nucleotide or amino acid sequence is known.

2 *Insertion into vector.* The vector is the replicon that will enable the gene to be maintained in the host cell and includes plasmids and phages for bacterial hosts. Plasmid vectors should have single sites for common restriction endonucleases and antibiotic resistance determinants that allow selection of transformants. The vector is cut with the same enzyme as that used to generate the chromosomal DNA fragments, and fragments and linearized vector are incubated with DNA ligase which covalently joins the DNA molecules. A heterogeneous population of molecules results, including dimers, trimers and multimers of fragment and recircularized plasmids. Some plasmids will contain an inserted fragment thus producing a hybrid, recombinant plasmid.

3 *Transformation of host cell.* The ligated plasmid mixture is introduced into bacterial cells specially treated so that they take up DNA in a process termed transformation. In most cases *E. coli* is the preferred host with the advantages that calcium chloride

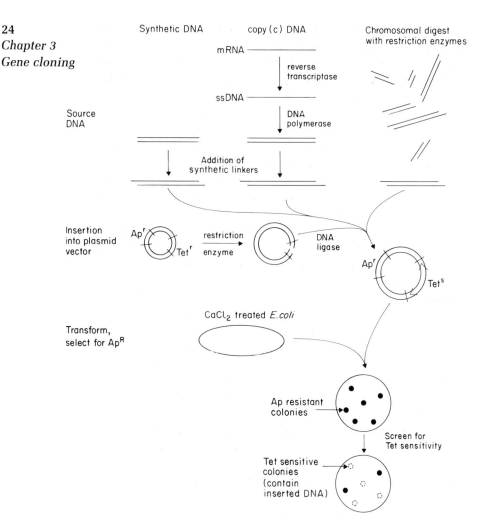

Fig. 3.1. Common strategies for cloning DNA in *E. coli*. Source DNA from bacteria is generally chromosomal DNA cut with a restriction endonuclease, from eukaryotes it is synthetic or cDNA. A synthetic linker sequence is attached to these latter fragments to provide the correct single stranded sequence to enable insertion into the vector. This is ligated into a suitable plasmid vector and transformed into *E. coli*. Ampicillin resistant colonies contain plasmid DNA and tetracycline sensitive colonies contain plasmid in which the tetracycline resistance gene has been inactivated by the insertion of foreign DNA. These clones can be examined for the presence of the desired gene as described in the text.

treated cells are highly transformable, the molecular biology of this bacterium is well understood and a variety of plasmid and phage vectors are available. Moreover, *E. coli* transcribes and translates most Gram-positive and Gram-negative genes with the exception of some actinomycete genes. Alternative hosts include pseudomonads, *Bacillus subtilis* and *Streptomyces*. *B. subtilis* has the advantages that it is nonpathogenic, has an efficient protein secreting system and is simple to grow on a commercial scale but

there are problems of instability of recombinant DNA in this host 25
and excessive proteolytic degradation of the product. Streptomy- *Chapter 3*
cetes should be particularly valuable for the cloning and expres- *Gene cloning*
sion of antibiotic synthesizing genes. Other hosts are being
developed at a rapid rate for specialist purposes such as
pseudomonads for biodegradation of hydrocarbons.

4 *Detection of the cloned gene.* In the example shown in Fig.
3.1 the insertion site within the vector is located within a tetracycline resistance gene. Transformants that are ampicillin resistant
but tetracycline sensitive therefore represent cells containing a
hybrid plasmid (insertional inactivation). Some common methods
for identification of the desired clone include (a) expression of
the gene and direct detection of the product (for example an
enzyme) or complementation of a mutation in the host. (b)
Immunological methods for screening of the desired product using
specific antisera. (c) Colony hybridization to a radioactively labelled probe DNA of the desired gene if the gene is unlikely to be
expressed. Once detected, the identity and structure of the gene
is confirmed by mapping and DNA sequence analysis.

5 *Maximizing expression of cloned genes.* In the simplest case,
the product is a protein derived from a single gene and the object
is to maximize yield. Nonprotein products resulting from a
metabolic pathway are more complicated to deal with and will
be briefly covered later.

For efficient transcription the cloned gene must be provided
with a powerful promoter preferably originating from the host
cell. Most plasmid vectors of *E. coli* use the *lac,* the *trp* or a hybrid
tac promoter as well as the phage lamba P_L promoter and in each
case the gene is inserted into a site such that transcription gives
rise to a fusion protein containing a few codons of the natural
gene from the promoter. Mutations of the *lac* operon that are no
longer susceptible to catabolite repression are generally used. The
operator region is retained however so that production of the
cloned gene product can be 'turned on' with inducer. It is generally
undesirable to have constitutive synthesis, particularly with toxic
proteins, since accumulation of the product may inhibit growth.
The *trp* promoter has the advantage over the *lac* promoter that it
needs fewer repressor molecules to inhibit transcription and control is therefore tighter. The hybrid 'tac' promoter comprises the
'−35 sequence from the *trp* promoter and the Pribnow box of
the *lac* promoter. Transcription is 11-times more efficient than
the catabolite-derepressed *lac* UV5 promoter and three times more
efficient than the *trp* promoter. The phage λ P_L promoter coupled
to a temperature-sensitive *cI* mutant repressor offers a strong
promoter under temperature sensitive control. Thus at 28°C transcription is undetectable but at 42°C the repressor is inactive and
up to 10% of the total cell protein derives from the cloned gene.

It is also useful to place an efficient termination signal after the gene so that readthrough into adjacent genes is prevented.

One of the great benefits of possessing a desired gene in cloned form is the opportunity for site directed mutagenesis. This is a highly specific and yet generally applicable form of *in vitro* mutagenesis that can be applied to a DNA molecule once the sequence has been determined. The process is outlined in Fig. 3.2. The DNA sequence to be mutated is inserted into M13, a single stranded DNA phage vector of *E. coli*. A synthetic oligonucleotide of 10–20 bases in length and containing the desired DNA sequence is annealed to the cloned gene for use as a primer. The second strand is synthesized using the Klenow fragment of DNA polymerase and the double stranded molecule is transformed into

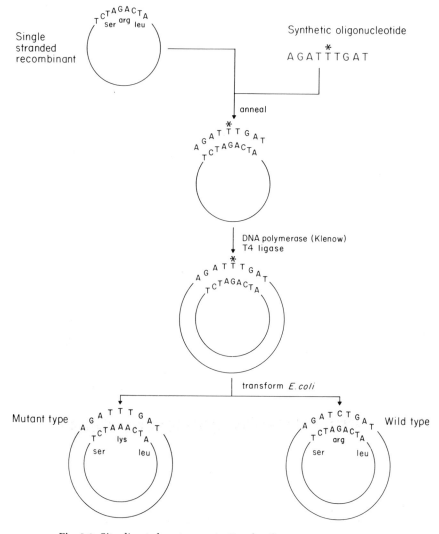

Fig. 3.2. Site-directed mutagenesis. For details see text.

E. coli. Some recipients will contain the mutated gene (Fig. 3.2) which can be recognized by phenotype or by using a ^{32}P labelled oligonucleotide probe. This procedure has been extremely valuable for eliciting predetermined changes of a single base in promoters or operators or of amino acids in signal sequences or the catalytic sites of enzymes.

The use of multicopy plasmids has two benefits; plasmid partitioning into daughter cells at cell division is more efficient and consequently plasmid maintenance more stable. Secondly, the gene dosage effect provides a higher concentration of mRNA and, since ribosome binding is most probably the rate-limiting step in protein synthesis, the provision of maximum copies of the message will be beneficial. It is useful to have the copy number of the plasmid under control so that the host cell is not continually burdened with 200 copies of the plasmid. The antibiotic resistance plasmid R1 naturally has a low copy number but temperature sensitive, 'runaway' mutants have been isolated that at high temperature accumulate up to 2000 copies per chromosome. Such derivatives of R1 constructed with unique restriction endonuclease sites under *lac* operator/promoter control have been developed and offer tremendous possibilities.

Efficient translation is also important. Promoters will contain information for the correct RBS in the message if the cloned gene is being synthesized as a fused protein. The SD sequence should be close to the consensus sequence of the host organism, particularly in *B. subtilis* where complementarity between the SD sequence and 16S rRNA is very important. In some instances the initiation codon will be that of the cloned gene and it is then necessary to optimize the distance between the SD sequence and the initiator codon. In the case of interferon genes inserted into plasmids containing the *trp* promoter and RBS, the optimal distance was nine nucleotides and using the *lac* promoter for expression of human growth hormone gene, 11 residues was most efficient. The base composition of the intervening sequence may also be important. In *B. subtilis* translation seems to be much more efficient with the bases A and T following the SD sequence than G and C.

Finally, the stability of the message is relevant and structural modification to the mRNA could result in greater resistance to nucleases. Double stranded RNA is more resistant to nuclease attack than single stranded and it may be that hairpin loop formation through the inclusion of inverted repeat regions at the 5' end of the mRNA preceding the SD sequence could provide additional stability to the molecule.

Secretion of proteins has the theoretical advantages of increased yield and ease of recovery and purification, and thus considerable effort has been devoted to the construction of vectors that export cloned gene products. This work has focussed on *B. subtilis,* an organism that secretes large quantities of extracellular enzymes and uses, in particular, the signal sequences from α-amylase of *B. amyloliquefaciens* and β-lactamase of *B. licheniformis.* By cloning a gene into a restriction site adjacent to the signal sequence the promoter and RBS of the vector are used to produce a fused protein bearing the signal peptide (Fig. 3.3). Secretion and processing (removal of signal peptide) of a β-lactamase lacking its own signal sequence has been obtained in this way as has secretion of human α2 interferon, and Semliki Forest virus glycoproteins E1 and E2. In both cases the genes were cloned into secretion vectors based on the α-amylase signal sequence and using *B. subtilis* as the host, however, the yields do not approach those of the α-amylase itself; for β-lactamase it was about 10% and for Semliki Forest glycoprotein E1 about 0.01% of the α-amylase. It appears that the genes were expressed efficiently, the low yield was either due to host exoproteases, or in the case of E1, inefficient secretion. In this connection, a great deal of effort is successfully being devoted to obtaining host strains of *B. subtilis* completely deficient in protease synthesis. Nevertheless, the potential for the synthesis and secretion of foreign proteins in *B. subtilis* has been clearly demonstrated.

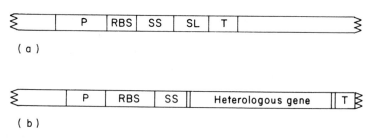

(a)

(b)

Fig. 3.3. Drawing of (a) part of a secretion vector and (b) the vector containing a cloned heterologous gene. P, promoter; RBS, ribosome binding site; SS, signal sequence; SL, synthetic linker containing several restriction enzyme sites; T, terminator. The cloned gene is designed to be expressed bearing a signal sequence that will effect secretion of the protein (*see* Doi 1984).

Much has been written about the value of obtaining high expression of cloned genes, but a less well understood problem pertains to vector stability. Obviously there is little to be gained from massive expression of a plasmid-borne gene if the plasmid places such a physiological burden on the host cell that it is rapidly lost from the culture and replaced by a smaller, deleted plasmid that poses less of a burden. Moreover, plasmid-free cells may predominate due to their faster growth rate. Deletions and rearrangements of recombinant plasmids during prolonged cultivation are referred to as structural instability and complete loss of plasmid as segregational instability.

Structural instability is difficult to overcome because it is not well understood. Deletions, insertions and rearrangements of DNA have been reported to occur in both *E. coli* and *B. subtilis* and are more prevalent in the latter. Both homologous recombination and insertion sequences have been implicated and although recombination deficient hosts minimize the problem these bacteria grow poorly. Perhaps the most successful route to alleviating structural instability is to place the gene under strict transcriptional control by way of a powerful regulated promoter such as the *lac* or *tac* promoters coupled with a regulated copy number (*see* above). With expression virtually eliminated during exponential growth the plasmid will be less of a burden to the cell. Integration of the cloned gene into the chromosome has been a successful stabilization strategy in *B. subtilis* where integration through homologous recombination can be readily induced (*see* Doi 1984).

Segregational instability occurs through defective partitioning of plasmid molecules into daughter cells at cell division followed by rapid growth of the plasmid-free cell population to the detriment of the slower growing, plasmid containing cells. High copy number plasmids are not very susceptible to this form of instability, since there is a high probability that at least one plasmid molecule will be distributed into daughter cells at each cell division. Low copy number plasmids, on the other hand cannot rely on probability to ensure accurate partitioning and contain a partitioning function (*par* region) which is responsible for distribution of daughter plasmids into daughter cells. Some vectors can be stabilized by cloning *par* regions from other plasmids into them. This is often insufficient for large-scale fermentations where other strategies have been developed. For example an essential gene can be placed in the vector, such as the valyl-tRNA synthetase (*valS*) gene, and a *valS* deficient host used. In this way recombinant plasmids have been stabilized for up to 150 generations in continuous culture (Nilson & Skogman 1986).

As noted above, it is pointless to devote effort to the efficient transcription and translation of a foreign protein if it is recognized as such by the host organism and degraded proteolytically. Resistance to host proteases is poorly understood but an ATP-dependant *lon* protease has been implicated in *E. coli*. There are two approaches to avoiding this problem. Protease deficient host strains (in particular *lon* protease negative strains) have been developed but it seems likely that strains completely lacking proteases will be non-viable. Alternatively the foreign proteins can be stabilized by fusing them to native proteins. This route was adopted for the production of somatostatin and human insulin in *E. coli*. A synthetic version of the somatostatin gene was cloned into the *lacZ* (β-galactosidase) gene in the plasmid vector pBR322 such that a β-galactosidase-somatostatin fusion product was synthesized. The synthetic gene was preceded by a methionine codon and, because somatostatin does not contain any internal methionine residues it could be cleaved from β-galactosidase by treatment with cyanogen bromide *in vitro*. The strategy is of little use for proteins that contain internal methionine residues since the product would be fragmented by the cyanogen bromide treatment. Specific proteolysis using trypsin has proved successful in some instances.

It has been noted that in many cases foreign proteins accumulate as aggregates or inclusions in the cytoplasm of *E. coli*. These inclusions offer some resistance to proteolysis but may be more problematical to extract and purify than soluble proteins, since denaturants such as urea or detergents must be used. However, this may not always be the case since high expression of the bovine growth hormone gene in *E. coli* resulted in crystalline cytoplasmic granules that could be easily purified from crude cell lysates by differential centrifugation and a simple washing procedure.

Three simple examples will be used to cover the remarkable achievements of recombinant DNA technology in bacterial strain improvement. The first example concerns gene cloning in *Streptomyces*. Genes for antibiotic synthesis pose problems because perhaps 10–30 enzymes are involved in a metabolic pathway. An early approach adopted by Hopwood's group was to identify and clone a key enzyme; p-amino benzoic acid synthetase (PABA-synthetase) in candicidin biosynthesis in *S. griseus*. There is a close correlation between PABA-synthetase activity and candicidin synthesis suggesting that this enzyme may be a controlling step in the synthesis of the antibiotic. The gene was cloned into *S. lividans* by inserting cleaved chromosomal DNA into a plasmid vector (pIJ41) and selecting for prototrophic transformants of a PABA-requiring auxotroph. Prototrophic clones contained a 4.5 kb fragment of *S. griseus* DNA which could be returned to *S. griseus* in a high copy number plasmid. This should increase the antibiotic yield by relieving a 'bottleneck' in the biosynthetic pathway.

As an extension of this, it has become evident that antibiotic biosynthetic genes are usually clustered. Moreover, the gene coding for resistance to the antibiotic normally resides within the cluster. Thus by cloning the resistance gene, most, and sometimes all, of the biosynthetic genes are simultaneously cloned. An example is actinorhodin, a polyketide antibiotic from *S. coelicolor* A3(2) which was cloned in a 25 kb segment of DNA although detection in this case was by complementation of a blocked mutant. The *act* cluster is expressed as several, in some cases polycistronic, transcripts and a putative activator gene was located centrally in the cluster. Multiple copies of this region led to over-production of actinorhodin.

The second example involves β-amylase synthesis by *B. stearothermophilus*. β-Amylases are typically plant enzymes and hydrolyse starch into β-D-maltose. They have commercial potential in the manufacture of maltose syrups but until recently no enzyme with sufficient temperature stability had been identified in a micro-organism. A strain of *B. stearothermophilus* was isolated which secreted a suitable enzyme but the yield was low, the organism grew poorly and traditional mutation programmes to increase yields were unsuccessful. By cloning the gene into *B. subtilis* using established techniques economic yields of the enzyme have now been prepared in pilot plant scale.

Finally, *Pseudomonas* species contain a variety of plasmids that code for enzymes that degrade a range of natural and synthetic organic compounds. These organisms are important for the natural degradation of toxic organic wastes in the environment. The hydrocarbons are generally readily degraded but halogenated compounds used in herbicides and insecticides are often resistant to attack. To broaden the substrate range of *Pseudomonas* strains, different plasmids were introduced into a host and following *in vivo* rearrangements, recombinant plasmids arose that had extended substrate specificities. For example, following the introduction of plasmids for catabolism of 3-chlorobenzoic acid and toluene into a strain of *P. cepacia*, recombinants that could also use 3–5 dichlorobenzoic acid and 4-chlorobenzoic acid were isolated. Using similar strategies, strains have been developed that effectively degrade 2,4,5, trichlorophenoxy acetic acid (the major, defoliating component of 'agent orange' used in the Vietnam War) in soils.

Gene cloning in *Saccharomyces cerevisiae*

Almost all yeast systems require the initial isolation of recombinant DNA to be carried out in *E. coli*. Two main types of vectors have been used; those which integrate into a yeast chromosome and those which (like bacterial plasmids) replicate autonomously. Both types carry genetic markers which can be selected in recipient organisms defective in that gene (for example the LEU2 and TRP1 genes coding for steps in the synthesis of leucine and tryptophan respectively and which are used to complement the DNA in mutants which require those amino acids). In addition both also must contain an origin of replication for maintenance in *E. coli* and carry single sites for one or more restriction endonucleases for the cloning of foreign DNA fragments. Perhaps the most popular vectors are the yeast episomal plasmids. Many strains of *Saccharomyces cerevisiae* contain a plasmid Scp1 (2μ) which replicates independently of the chromosomes. It is usually present at 50–100 copies per cell and is of known nucleotide sequence. Cloning vectors containing the replication origin of this plasmid behave as autonomously replicating units and transform yeasts with high efficiency. Chimaeric plasmids containing the origin of replication of the 2μ plasmid plus an *E. coli* origin act as 'shuttle vectors' which may be used both in *Saccharomyces cerevisiae* and in *E. coli.* As in *E. coli,* a strong promoter is necessary for a high level of gene expression and control over the expression system is desirable. These promoters include those for alcohol dehydrogenase (ADH1) and 3-phosphoglycerate kinase (PGK). Transformation is more difficult to achieve in yeast than in *E. coli.* The usual method employed involves removing the cell wall enzymically (with glusulase, helicase, zymolase, lyticase etc.) and then transforming the sphaeroplasts produced in the presence of polyethylene glycol. The sphaeroplasts are then allowed to regenerate in selective solid media. In addition to applications to the fermentation industries, yeast systems have been used for the experimental production of a number of mammalian proteins including interferon, human serum albumin, insulin-like growth factor, epidermal growth factor, and the hepatitis B virus surface antigen.

Chapter 4. Animal cell cultures

Most animal cells can be grown *in vitro* using synthetic medium supplemented with serum or some other complex nutrient source. Tissue culture strictly refers to tissues grown *in vitro* but is generally used to include the culture of dissociated cells or cell culture. Primary cell lines comprise tissue explants treated with a protease such as trypsin to dissociate the cells and cultured in Petri dishes containing a suitable nutrient medium. Human cell cultures established in this way generally die after about 50 generations, chicken cells in a shorter time and mouse cells rapidly become moribund.

Established cell lines have been selected to survive for extensive periods *in vitro*. Common examples include L and 3T3 cells from mouse embryo, BHK from baby hamster kidney, CHO from Chinese hamster ovary and human fibroblast cell lines. Many of these established lines will survive for months or years so long as they are recultured at frequent intervals in much the same way as a microbial culture is subcultured.

Immortal cell lines are generally of malignant origin or are normal cells transformed by a virus such as the Epstein Barr virus. The most famous example of an immortal cell line originated from a cervical carcinoma from *He*len *La*ne and it has been speculated that at any one time there are some 17 tonnes of HeLa cells worldwide! Transformed cell lines grow indefinitely and to a higher cell density than normal cells. Unlike normal cells, they have no requirement for a solid surface on which to adhere (anchorage dependance) and they have a lower serum requirement for growth. However, they may be of limited biotechnological value since they are of malignant origin and may release transforming viruses which could contaminate the product.

Animal cell cultures have been used for several decades for the propagation of viruses for use in vaccine preparation but recent developments have enabled their exploitation for the synthesis of specific protein products including monoclonal antibodies and with advances in genetic engineering there is potential for both cell cultures and complete animals to be genetically modified for new products and increased expression of existing products.

Products from animal cell cultures

Many people are now realizing that bacteria and yeast may not be the best systems for manufacturing complex mammalian proteins such as hormones and blood factors and a more 'natural' source, cultured animal cells, may be more suitable. In particular, animals cells should be capable of correctly processing the protein and secreting it into the culture medium where appropriate. This is especially important for proteins such as fertility hormones which may be 30% carbohydrate and bacteria lack the necessary glycosylation enzymes. There are problems however relating to slow growth rate, the low cell densities achievable *in vitro* and low product yields.

Some animal cells are anchorage-dependent and grow only when attached to a solid substrate while others may be grown as discrete cells in agitated suspension cultures in systems similar to those used for micro-organisms. This separation is not absolute, however, as many cell lines will grow in suspension culture at a much lower productivity of cell mass or product than when grown on a substrate. In addition, suspension-style cultures for anchorage-dependent cells may be created by the use of microcarriers.

Irrespective of the cell line, all animal cell cultures require an adequate supply of oxygen and tolerate only modest changes in pH. Being heterotrophic, they also require a carbon and energy source (usually glucose), together with amino acids, trace metals and specific growth factors and hormones. Traditionally, many of the latter requirements were met by the inclusion of (fetal) calf serum in growth media but problems of cost, supply and the inherent variability of this material has led to attempts to use 'serum-free media', especially in large scale growth systems. Growth factors are polypeptides which control the growth of normal cells and are coded by cellular genes (oncogenes). Over 40 different growth factors have been identified and include platelet-derived growth factor, epidermal-growth factor, the insulin-like growth factors, interleukin-2 and nerve growth factor. There is no serum free medium suitable for all cell lines since many have specific growth requirements and some workers prefer to add small concentrations of serum ($< 1\%$ v/v) to ensure optimal growth.

The cell line most acceptable for the production of many animal cell products is the human diploid fibroblast. This is a nontransformed and genetically stable cell line which will not grow in suspension culture. Unfortunately it exhibits only low productivity, requires serum and has a limited life-span. The initial scale up of such anchorage-dependent cells employed a multiplicity of small culture units such as flasks and roller bottles. These have been replaced to some extent by a variety of systems including stack plates, multitrays, high surface area matrices, immobilized beds employing glass beads and microcarriers. Beads based on cross-linked dextran are probably the most widely used microcarriers although others employing cross-linked gelatin, polystyrene or polyacrylamide are also available commercially. Dextran-based microcarriers have been used successfully for the growth of human fibroblasts at volumes of several thousand litres, for example in the production of interferon B.

Large scale growth of non anchorage-dependent cell lines has been practised for some years. For example the manufacture of foot and mouth virus vaccine employs baby hamster kidney cells in 3000 l stirred tank reactors while lymphoblastoid interferon has been produced on a 8000 l scale. Hybridoma cell lines producing monoclonal antibodies (MAB) are not anchorage dependent. Small scale production (up to gram quantities of MAB), may be carried out by growing the cells *in vivo* in rats or mice as ascites tumours. While the yields of MAB may be high, very large numbers of animals must be sacrificed to produce large quantities of antibodies. The increased requirement for MABs in such diverse applications as diagnostic reagents, tumour imaging, immunopurification technology and potentially also for therapy on a large scale has led to the development of a number of methods for the large scale *in vitro* culture of hybridomas. Methods employed include the use of homogeneous suspension cultures in airlift and stirred tank reactors and cells immobilized or entrapped in hollow fibre perfusion systems, microcapsules, agarose microbeads and ceramic cartridges.

The 'strain development' aspect of animal cell cultures is progressing rapidly and three systems are particularly noteworthy (Fig. 4.1).

The cellular enhancer element is a short nucleotide sequence that directs lymphocytes to produce large amounts of antibody molecules. When placed less than one kilobase from other gene sequences it will 'enhance' the expression of these genes. There seems to be little specificity for the nature of the gene and by placing both enhancer and product gene in a suitable vector, high expression of several genes has been achieved. The enhancer sequence is specific for lymphoid cells however and does not operate in other cell types. Similarly, several viruses including SV40, Maloney Sarcoma virus and Rous Sarcoma virus have repeat sequences upstream which are required for the expression of viral genes and can be used to enhance expression of cloned genes such as β-globulin.

A second expression system for animal cells uses the bovine papilloma virus (BPV) cloning vector. This exists as a multicopy plasmid and therefore provides high yield through gene dosage. Vectors comprising portions of the BPV genome linked to a mouse metallothionein promoter have been constructed. Metallothioneins are small, cysteine-rich, heavy metal-binding proteins that are induced by cadmium, mercury or other heavy-metals. By inserting various genes adjacent to this promoter, regulated syn-

THREE MAMMALIAN GENE EXPRESSION SYSTEMS

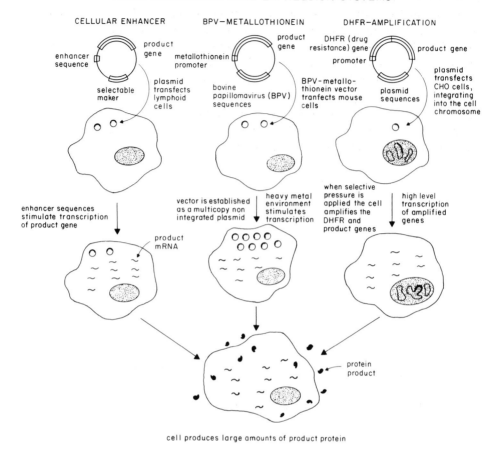

Fig. 4.1. Strategies for obtaining expression of foreign genes in mammalian cells.
(Adapted with permission from Wilson 1984.)

thesis of product by adding cadmium to the medium has been achieved. Tissue plasminogen activator and leutinizing hormone genes have been expressed in this way.

The third system involves CHO cells which, when subjected to increasing concentrations of the toxic drug methotrexate, amplify the gene for dihydrofolate reductase (DHFR). By placing the gene of interest next to the dehydrofolate reductase sequence, both genes are amplified under selective pressure. High yields of several proteins have been achieved in this way, but as in all these systems problems of product stability, correct processing and efficient secretion, particularly of foreign proteins have yet to be completely solved.

Monoclonal antibodies

One animal cell product that is attracting ever-increasing attention is the monoclonal antibody. This is a specific antibody produced from a normally short-lived, antigen-activated B cell that has been immortalized by hybridizing it with a myeloma cell. Thus the hybrid retains the ability of the B cell to secrete antibody and the ability of the myeloma cell to grow indefinitely. The advantage of the monoclonal antibody is that it is derived from a single cell and comprises a uniform breed of antibody specific for a single antigen site (epitope). Traditional, polyclonal antisera are derived from many cells and contain heterogeneous antibodies that are specific for all the epitopes in an antigen (Fig. 4.2).

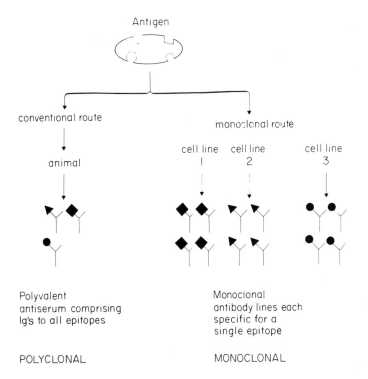

Fig. 4.2. A comparison of the specificities of polyvalent antisera and monoclonal antibodies. Polyvalent sera contain immunoglobulins against all the antigenic sites (epitopes) in the antigen, monoclonal antibodies are derived from a single lymphocyte and are specific for one epitope.

The preparation of monoclonal antibodies is outlined in Fig. 4.3. A mouse is immunized with the desired antigen to generate activated B cells. When a high titre has been achieved the spleen is removed and the lymphocytes are fused with myeloma cells by incubating the mixture in the presence of polyethylene glycol. At best only about 1 myeloma cell in 10^3 fuses with a spleen cell and it is therefore necessary to select against the unfused cells. B lymphocytes die *in vitro* and to select against myeloma cells, lines defective in hypoxanthine phosphoriboxyl transferase (HPRT) or

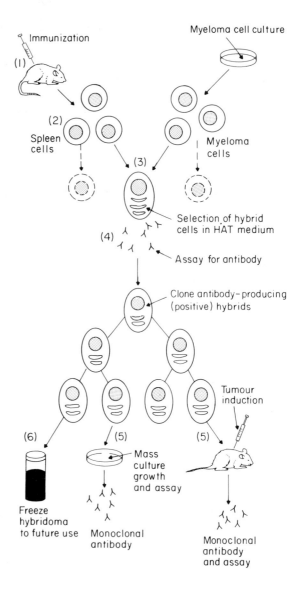

Fig. 4.3. Principal steps involved in the production of monoclonal antibodies (see text for details).

thymidine kinase (TK) are used. HPRT and TK are salvage enzymes that convert hypoxanthine or thymidine into usuable nucleotides. HPRT- and TK-deficient strains are unable to use hypoxanthine and thymidine respectively and rely on *de novo* synthesis of nucleotides the pathways for which are sensitive to aminopterin (A). Thus the addition of aminopterin (in the presence of hypoxanthine and thymidine; HAT selection) kills the original myeloma cells and selects for hybrids that have received the HPRT (or TK) gene from the spleen cells. The hybrids are then 'cloned' such that each well of a microtitre plate receives one or a few

clones and the hybrids are screened for the desired antibody.
Positive clones are subcloned by dilution and redistribution in
microtitre plates and positive clones characterized. The amount,
specificity, binding affinity and nature of the antibody (isotypes
of the heavy and light chains) together with the ability of the
hybridoma to produce consistent amounts of antibody are of
concern.

As indicated earlier, production of monoclonal antibodies
can be achieved *in vitro* but the yield is often low (about 10 μg/ml
specific antibody). Large scale culture in airlift fermenters is now
a commercial reality and can provide higher yields of some 50
μg/ml. A convenient and economic way to produce fairly large
amounts of antibodies is to grow the hybridomas in histocompat-
ible or immunocompromised mice as ascites tumours. A mouse
bearing an ascitic tumour may accumulate a few millilitres of
ascitic fluid which may contain 5–20 mg/ml specific antibody.

Gene cloning in animal cells

As with gene cloning in bacteria, the propagation of foreign genes
in animal cells required methods for introducing exogenous DNA
into the cell and vectors for its stable replication and maintenance.

DNA-mediated gene transfer can be achieved by indirect or
direct methods. Indirect methods are generally used with culture
cells and the most practical has been the DNA phosphate precipitation method in which DNA is coprecipitated with calcium phosphate to form microscopic aggregates. These bind to cell surfaces
and are ingested endocytotically. Within 7 h DNA/calcium phosphate can generally be detected in the nucleus and is expressed
after 48–84 h. Direct methods involve microinjection into the
nucleus of a somatic cell or a zygote. In both instances the DNA
finds its way to the nucleus where it is readily expressed in a
transient fashion and homologous recombination can give rise to
integration of the gene.

An early example of gene transfer involved the transfection of
the thymidine kinase gene from herpes simplex virus into mouse
cells that were deficient in the enzyme. The powerful HAT selection mentioned earlier could be used to select for cells that received
the gene. Nonselectable genes can be readily introduced into cells
by either ligated or nonligated cotransfer in which the gene is
either ligated to or simply mixed with a selectable gene (for
example TK) and a large proportion (as high as 80%) of the cells
containing the selected gene also contain the nonselected one.

Gene transfer systems are proving to be invaluable for the
isolation of various genes. In one general strategy (Fig. 4.4) rodent
cells are transfected with total human DNA and the appropriate
marker is selected. DNA from these primary transformants is used

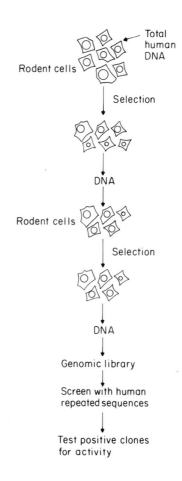

Fig. 4.4. Strategy for gene isolation by gene transfer techniques. This protocol may be used to isolate genes whose products can be selected or easily screened for in somatic cells. (From Kucherlapati 1984 with permission).

for successive rounds of transfection to 'dilute out' the human DNA. A genomic library prepared in bacteriophage λ is then screened with human repetitive sequence probes. The human DNA segments can be identified since they carry flanking repeated sequence (*Alu*) DNA. Positive clones are then tested for activity. In this way several human oncogenes have been isolated.

In many instances it is desirable to have the transfected gene situated in a vector rather than integrated into the chromosome. Many viral vectors have been developed based on SV40, adenovirus, bovine papilloma virus, retroviruses and vaccinia virus. Of these, just three will be mentioned briefly here; SV40, BPV and vaccinia vectors.

Simian virus 40 is a small virus with a double strand DNA genome of 5.3 kb. Since much is known of the molecular biology of this virus it is a strong candidate as a vector for foreign DNA in mammalian cells. The virus has two modes of infection. In permissive cells, both early and late genes are expressed and it behaves as a lytic virus. In nonpermissive cells only the early gene is expressed and the viral DNA is integrated into the cell chromosome. These transformed cells adopt many of the characteristics of neoplastic cells. Due to the small size of the SV40 virus the amount of recombinant DNA that can be packaged is severely limited. To overcome this problem a 'helper' virus system has been developed in which essential late genes in the cloning virus are replaced with recombinant DNA to provide a defective virus. Monkey cells are cotransfected with this virus and a helper virus which has a temperature sensitive mutation in the early gene. Plaques at the non-permissive temperature must represent complementation of a recombinant (no functional late gene) virus with a helper (no functional early gene) virus. More advanced SV40 vectors employing regulated promoters, selectable markers and hybrid shuttle plasmids derived from SV40 and pBR322 and able to replicate in *E. coli* have been developed.

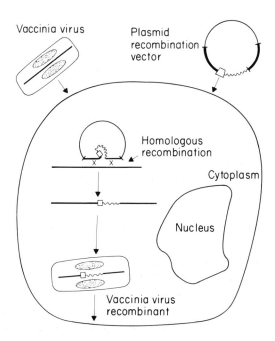

Fig. 4.5. Formation of vaccinia virus recombinants. Cells are infected with vaccinia virus and transfected with a plasmid recombination vector. These plasmids contain a foreign protein-coding sequence (⁓) fused to a correctly orientated vaccinia promoter (—▢—) and flanked by nonessential vaccinia virus DNA (heavy line). Within these infected cells homologous recombination between the virus genome and plasmid DNA results in insertion of the foreign gene into the vaccinia virus genome. The recombinant genome is replicated and packaged into infectious progeny virus. Total progeny virus from transfected cells is screened for virus recombinants as described in the text. (From Smith *et al.* 1984, with permission.)

Vectors based on bovine papilloma virus (BPV) are a recent innovation and hold great promise since they are extrachromosomal with a copy number of about 100. These viruses responsible for warts and other benign tumours can stably transform cultured monkey and mouse cells amongst others. A fragment (69%) of the genome that encodes the transforming ability has been identified and used as a vector to clone and express the rat-preproinsulin and human growth hormone genes.

Finally, vaccinia virus is a large virus that replicates in the cytoplasm of the cell and is used to vaccinate against smallpox since it is serologically very similar to the smallpox virus but avirulent. The double-strained DNA chromosome is large (187 kb) and could code for some 200 polypeptides. Nonessential areas of DNA have been identified which can be deleted or interrupted by insertions without effect on viral replication in tissue culture. There is great flexibility in the amount of DNA that can be packaged ranging from deletions of 9 kb to additions of 25 kb. There are three aspects of this virus to consider in the preparation of recombinants. The genome is so large that it cannot be handled easily *in vitro.* Second, isolated DNA is not infectious; enzymes associated with the virian are needed for protein synthesis and replication, so the recombinant DNA cannot be transfected into cells. Thirdly, vaccinia virus RNA polymerase recognizes only its own control signals which must be provided for the foreign DNA. A scheme for overcoming these problems is shown in Fig. 4.5 in which the cloned gene is placed in a plasmid vector and cotransfected into a cell with vaccinia virus. *In vivo* recombination provides a recombinant vaccinia virus.

Chapter 5. Plant cell systems

The biotechnological exploitation of higher plants centres on plant tissue culture techniques initiated by Haberlandt at the turn of the century. Today, the unique regeneration properties of plants and their biochemical potential are used in essentially three ways; *in vitro* propagation (i.e. micropropagation using tissue culture techniques) of plants, mainly in horticulture and forestry; the production of secondary metabolites from mass plant cell cultures and thirdly, the use of recombinant DNA techniques to modify plants genetically, particularly agricultural crops.

In vitro propagation of plants

Stem tips as long as 5–10 mm have been used, for example by orchid growers, to propagate plants since the 1950s. More recent developments involve the propagation of tiny (0.1 mm) apical shoot tips taken from the plants fast-growing meristem region. This meristem tissue is multiplying so rapidly that even in infected plants it is virus-free. This therefore permits the propagation of disease-free stock.

The basic procedure of micropropagation is shown in Fig. 5.1 and involves three stages: (1) selection and sterilization of suitable explants and their transfer to a nutrient medium; (2) proliferation of shoots in multiplication medium; and (3) transfer of the shoots to a rooting medium and planting out. These aspects will be outlined here. There are several approaches to micropropagation depending on the type of material and multiplication process used.

Meristem cells from auxillary shoots can be excised, usually sterile, and grown on a suitable solidified nutrient medium. The hormonal balance in this medium is important and generally contains high levels of cytokinin relative to auxin. This promotes shoot growth, the inverse proportions tend to promote root growth. By keeping the cytokinin concentration high, precocious branching of the explant is achieved with clusters of shoots. These can be subdivided and recultured to provide yet more shoots in a process that can be prolonged indefinitely. This technique is applicable to many woody plants in particular forest and orchard trees. Alternatively apical shoots or apical meristems can be used as the explants as noted above.

Multiplication can also be achieved by adventitious shoots (those shoots that arise other than at the normal leaf axil region)

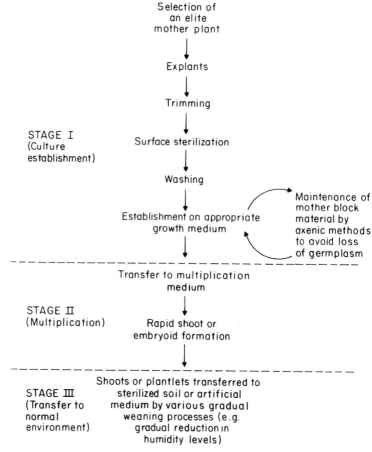

Fig. 5.1. Micropropagation of plants.

particularly amongst bulbs, corms and tubers. Shoot bases will regenerate clusters of shoots which can be subdivided as described above.

Finally callus tissue can be used for micropropagation. Numerous species form callus material when grown in tissue culture. This undifferentiated mass of cells can be induced to form shoot buds by careful balance of the medium. The buds are repeatedly removed and planted in solid medium until shoots and roots appear. As with the other micropropagated plants these are then planted out and 'hardened' off. In these ways disease-free stocks of hundreds of plants can be accumulated in months rather than years. This technique has proved to be successful in the mass propagation of the oil palm.

Mass plant cell culture

Plants are a particularly valuable source of pharmaceuticals and food additives including deigoxin, opiates, antitumour products and alkaloids. Natural product synthesis suffers the disadvantages of extremes of climate, diseases and pests, seasonal fluctuations and inconsistent product yield and quality. *In vitro* plant cell culture should overcome these problems. Plant cell technology, like microbial technology requires an inoculum, media, strain development to maximize yields and largescale culture systems.

Plant cells are totipotent and therefore any cell should be capable of producing the desired product typical of the plant. It is sensible to initiate cultures from high-yielding plants in an attempt to retain the productivity *in vitro* but the type of cell (i.e. one actively synthesizing product) seems less important. Obviously healthy, disease-free tissue should be used and carefully sterilized.

Medium composition is a critical feature of biomass and product yields. Typical media are similar to microbiological media and include salts (K^+ Na^+ Cl^- PO_4^{3-}), and nitrogen (NH_4^+, or glutamate), trace elements and growth factors (often in the form of coconut milk). Biomass yields of 5–20 g dry weight per litre are typical, but it must be remembered that many desired products are secondary metabolites and only synthesized by resting or differentiated cells. Therefore high growth rates may not necessarily be desirable (see earlier discussion on catabolite repression in bacteria) and indeed immobilization of resting plant cells is being explored for the efficient production of secondary metabolites. The usual carbon source is sucrose although other carbohydrates have been examined and glucose may often provide fastest growth rates and maximum biomass accumulation but may inhibit product yield (*see* comments above). The nature and concentration of the nitrogen and phosphate sources may have profound effects on product yield and the complement of growth factors is very important. The physical environment (light, temperature, pH, etc.) is also important. Optimal conditions for product yield are usually determined empirically by evaluation of these various factors.

Early (pre 1970) attempts to obtain secondary metabolites from plant cell culture yielded disappointingly low amounts of product. This situation has been markedly improved and *in vitro* yields of some materials far exceed parent plant material. To date, however, the only commercial success is the production of shikonin derivatives.

The selection strategies for high-yielding cell lines follow those used in traditional microbial strain development programmes. Populations of clones are screened for high product yield, often using radioimmune assay or some other sensitive technique. Mutagenesis both physical and chemical has been of controversial success in the development of high-yielding cell lines. The problems with these procedures include difficulties associated with cell cloning and slow growth rates but progress is being achieved. Recombinant DNA techniques are at an early stage in plant cell technology but as we understand more about the plant genome the approaches adopted by microbial molecular biologists such as exploitation of strong promoters, removal of catabolite repression will become realistic. In the near future we might expect product yield to be improved by gene dosage through cloning genes onto multicopy vectors but most products are the result of metabolic pathways involving many genes which complicates the genetic analysis.

Gene cloning in plants

The tremendous possibilities for improving agricultural crop plants using recombinant DNA technology required the development of suitable vectors analogous to the plasmid and phage vectors of bacteria. Plant viruses have some potential in this area. More than 90% of plant viruses have single stranded RNA genomes and have been little exploited as possible vectors. The caulimoviruses are the only group of plant viruses that have double-stranded DNA genomes and these have several shortcomings as recombinant DNA vectors. The host range of the 12 viruses in the group is limited; none has yet been isolated which will infect legumes or monocotyledons. They are pathogenic and it is not known if the virus DNA integrates into the host chromosome. Although these viruses are being developed as vector systems much work remains to be done.

The most promising plant vector system uses a bacterial plasmid (Fig. 5.2). The soil-borne, Gram-negative bacterium, *Agrobacterium tumefaciens* produces a neoplastic disease in many dicotyledenous plants; the Crown Gall tumour. Virulent strains of *A. tumefaciens* harbour a large (150–200 Megadalton) plasmid, the Ti-plasmid, which is responsible for induction of the tumour. This plasmid codes for enzymes that are responsible for the biosynthesis and degradation of one of a range of amino acid derivatives termed opines. Common examples include octopine and nopaline and are the basis of this highly developed parasitic interaction. Opines are not normally encountered in plant tissues and can be used as a sole source of carbon, nitrogen and energy by Ti-plasmid-bearing strains of *A. tumefaciens* thus allowing the growth and proliferation of the infecting bacterium.

Upon infection of a plant, generally through a wound, a segment (T-DNA) of the Ti-plasmid (about 10%) is integrated into the plant chromosome. The DNA is integrated at different sites in different tumour lines and codes for the opine-synthase and seven other transcripts. The ability to catabolize the opine is retained by the plasmid in the infecting bacterium. Some of the genes in the T-DNA are responsible for altering the levels of hormones in

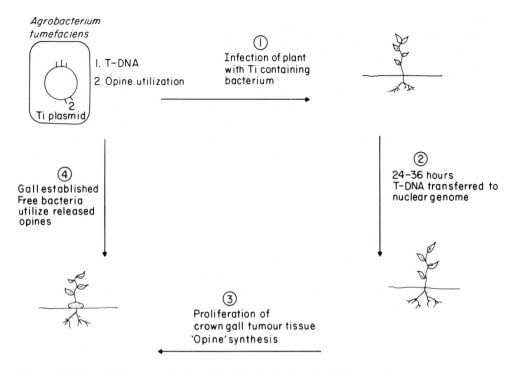

Fig. 5.2. Outline of plant gene cloning using *Agrobacterium tumefaciens*.

the plant tissue by affecting the ratio of auxin to cytokinin. This promotes the tumour, the morphology of which is partly dependent on the actual ratio of hormones. Thus tumour cells do not need continuing infection by *A. tumefaciens* to maintain the neoplasm, the T-DNA is sufficient and tumour cells will grow in tissue culture without supplementary auxin or cytokinin.

Since T-DNA contains genes that are expressed in plant cells, agrobacteria perform a true genetic manipulation of plants and contain all the functions necessary for the transfer, stable incorporation and expression of genetic information in this background. It should therefore be possible to exploit this system for the introduction of new genes into plants. Plasmids have been constructed that are mutated in their carcinogenic properties yet retain genes responsible for transfer of the T-DNA into the plant cells. One of these is an octopine-plasmid with transposon inactivation of the 'auxin-like' functions of the T-DNA. Tumours induced on tobacco plants with this plasmid produce numerous shoots. Most of these shoots upon separation from the host develop roots and grow into normal healthy plants with no detectable T-DNA. However, careful screening indicated a minority of plants containing and expressing the gene for octopine synthesis. One of them was studied in detail and regenerated into a morphologically normal plant with the octopine-synthesizing activity in all cells examined.

Furthermore, a nopaline plasmid has been used to clone a

yeast alcohol dehydrogenase gene in tobacco using a similar procedure. Cells from the tumourous growth could be regenerated into healthy tobacco plants which contained copies of the yeast gene in all cells examined. The gene was not expressed, however, and current activity is devoted to harnessing plant and T-DNA promoters for the efficient expression of foreign DNA. By cloning the gene for β-galactosidase into the nopaline synthase gene to construct an 'in frame' fusion, constitutive expression of an active β-galactosidase was obtained in sunflower and tobacco cells and similar strategies have been used for the expression of bacterial antibiotic resistance genes.

Chapter 6. Growth and fermentation systems

Fermenter construction

Fermenters used in the production of alcoholic drinks are traditionally of simple design: open rectangular or cylindrical vessels sufficed until recent years. Most other fermentation processes require enclosed vessels capable of operation under pure-culture conditions. Figure 6.1 shows a typical enclosed stirred cylindrical fermentation vessel.

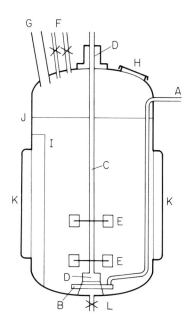

Fig. 6.1. Typical aerated stirred fermentation vessel.
(A) air main, (B) air sparger, (C) stirrer shaft, (D) shaft bearings, (E) stirrer paddles, (F) alkali, antifoam inlets, (G) exhaust, (H) manhole (for cleaning, addition of medium, etc.), (I) baffle (one of 3 or 4), (J) static medium level, (K) attemperation jacket, (L) harvest.

The vessel is constructed of a suitable grade of stainless steel, with butt-welded joints polished flat on the inner surfaces. Overlapping joints are unacceptable since the resulting ledge causes an accumulation of spent culture, creating problems on subsequent cleaning and sterilization: the high concentration of residual micro-organisms requires longer heat-treatment to sterilize.

Actively metabolizing cultures generate heat; attemperation of fermentation vessels is essential. Smaller vessels, up to 1000 litres, can be controlled effectively by a cooling jacket. With larger vessels the jacket/fermenter wall surface may be insufficient for effective cooling, and an internal cooling coil is necessary. However, a cooling coil, by providing horizontal surfaces within the vessel, is an additional complication to cleaning procedures. In modern equipment, cleaning is normally performed automatically by spray jets but it may be impossible to locate jets to clean all surfaces of a complex cooling coil system, and manual cleaning is then necessary.

Most fermenters are fitted with stirring gear, which is particularly necessary to break up clumps of mycelial organisms. Various designs of agitator have been used, but the commonest is a set of flat vertical paddles set in a horizontal disc, to create a circular movement of the medium. The tendency for vortex formation is prevented by vertical baffles fitted to the vessel walls. The bearing of the stirrer shaft has to support the mechanical stress of stirring a large volume of viscous culture, but yet prevent access of micro-organisms which might be drawn into the vessel by the rotation of the shaft. In the larger sizes of vessel, a continuously steam-sterilized section of the bearing prevents such contamination, with the additional benefit of lubrication by sterile condensate.

Although some fermentations operate anaerobically, the great majority are aerobic processes requiring large volumes of sterile air. Air is supplied through a suitable sparger: a multihole sparger in the form of a ring or grid (Fig. 6.2) is effective with unicellular organisms and, indeed, the rising bubbles of air may provide all the mixing required. Multicellular organisms tend to overgrow most of the holes of ring or grid spargers and a single open pipe below the lowest stirrer is then more effective. The large diameter

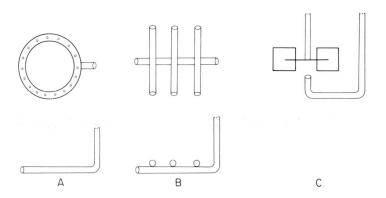

Fig. 6.2. Spargers for fermentation vessels. Ring sparger (A), Grid sparger (B), single open pipe (feeding immediately below lowest stirrer paddle) (C).

pipe cannot be blocked by mycelial growth, and the large bubbles are quickly broken and dispersed by the stirrer.

Aeration and agitation

In addition to providing oxygen, aeration is also important in purging the culture of unwanted volatile products of metabolism. Agitation either by stirring or as a side effect of aeration, is necessary for five reasons in particular:

1 To increase the rate of oxygen transfer from the air bubbles to the liquid medium; micro-organisms cannot use gaseous oxygen, but only oxygen in solution.
2 To increase the rate of oxygen and nutrient transfer from the medium to the cells. By continuous movement, cells are prevented from creating stagnant areas of low levels of nutrient and oxygen.
3 To prevent formation of clumps of cells, or aggregates of mycelium.
4 To increase the rate of transfer of products of metabolism from cells to medium.
5 To increase the rate or efficiency of heat transfer between the medium and the cooling surfaces of the fermenter.

All of these effects are improved by movement of the medium, but in addition the turbulence of agitation improves aeration in the following ways:

1 by dispersing the air in smaller bubbles;
2 by causing the bubbles to follow a more tortuous path and delaying their escape from the culture;
3 by preventing the coalescence of bubbles;
4 by decreasing the rate-limiting thickness of the liquid film at the gas/liquid interface.

The optimal arrangement of the stirring and aeration equipment has to be determined for each different type of fermentation by pilot plant studies. In aerobic cultures, the supply of oxygen to cells should not be the rate-limiting factor: that is an inefficient (and therefore uneconomic) operation.

The following are particularly important in affecting the supply of oxygen to the culture:

1 *Type of agitation:* the shape, number and arrangement of impellers and baffles. A large fermenter is normally fitted with two or three impellers at suitable levels on the stirrer shaft, and either three or four vertical baffles on the wall of the vessel.
2 *Speed of agitation.* Laboratory fermenters may operate at 1000 rpm or higher, but this is impracticable with large vessels. The consistency of the culture in a penicillin fermentation, for example,

is such that even 50 rpm may require an uneconomically high input of energy. In practice, the shear effect at the tip of the paddle is the important factor, and tip speed rather than rpm is the important figure.

3 *Depth of liquid in the fermenter.* Bubbles remain longer in the medium of a tall, or deep, fermenter; and the greater hydrostatic pressure at the sparger improves solution of oxygen. In practice a height:diameter aspect ratio of 3:1 or 4:1 is common.

4 *Type of sparger* (*see* above) is a compromise between numerous easily-blocked holes and one single opening producing very large bubbles. Preliminary testing is necessary to develop the best design for each type of fermentation.

5 *Air flow.* Aeration efficiency is increased by increasing the air flow rate, normally expressed in terms of vvm (vol of air/vol of medium/min). In practice, large fermenters cannot be supplied economically with air at greater rates than 0.5–1.0 vvm.

6 *Physical properties of the medium:* temperature, viscosity, surface tension and the nature of the organism all affect solution of oxygen directly or by bubble size and turbulence. In practice these effects are not alterable, being dictated by microbiological requirements. Increased pressure improves solution of oxygen and can be applied either by designing a deeper (taller) fermenter or by partially restricting the escape of exhaust gas to increase the top pressure.

Oxygen is normally supplied as air. While it is theoretically possible to improve aeration efficiency by supplying pure oxygen from cylinders, this is very seldom done. The air supply is, usually sterilized by filtration, heat-sterilization being uneconomic. Adiabatic heating in the compressors does substantially reduce the microbial population of the air; not, unfortunately, sufficiently for sterilization. Formerly carbon was a common filter material, but now mineral wool filters, impregnated with polystyrene resin to prevent channelling, are most commonly used.

Small fermenters are normally fitted with individual air pumps, flow meters and filters (Fig. 6.3a). Note the situation of the air flow meter in the nonsterile section before the filter, to permit removal if necessary for maintenance without contamination of the system. Large installations (Fig. 6.3b) are fitted with a bank of compressors feeding a bank of air filters which in turn supply a sterile air main to all fermenters and often, also, a supply of pneumatic power for automatic valves, etc.

Sterilization of fermenters

In the laboratory, the standard method of sterilization is heat: 120°C moist heat for 15 min, or 160° dry heat for at least 1 h. On an industrial scale, dry heat is prohibitively expensive for all but

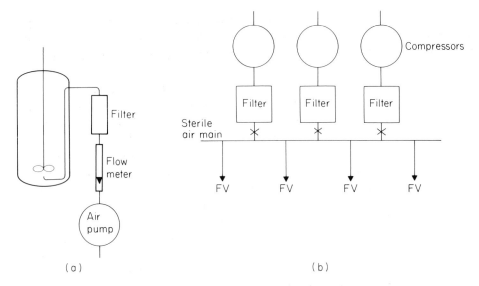

Fig. 6.3. Air supply to fermenters.
(a) Laboratory or pilot scale.
(b) Plant scale (FV: fermentation vessel).

a few specialized purposes (for example incineration of exhaust air from culture of pathogens used for vaccine preparation) and moist heat sterilization is the customary method employed.

The laboratory sterilization regime of 120°C for 15 min does not guarantee sterility, but with volumes of medium of 1 litre or less, and normal levels of contamination of the ingredients, the probability of an infected batch is less than 1 in 1000. As the scale of operation increases, sterilization must be more rigorous, by increasing time or temperature, or both, to maintain the generally acceptable probability of contamination (i.e. one surviving cell or spore) at 1 in 1000.

Sterilization of a fermentation vessel may take place with a full charge of medium, added nonsterile, or the sterile medium may be added after sterilization of the empty fermenter. Another common method, used to avoid damage to the nutritive value of the medium by caramelization reactions, is sterilization of the sugar component of the medium separately and adding to the other nutrients which were sterilized in the fermenter. In modern fermentation equipment the processes of sterilization and cooling are under automatic control; the sequence of operations is as follows:

Sterilization of fermenter and medium together

The ingredients of the medium, 'mash', are added through the manhole in the top of the vessel or pumped in from a separate mixer unit and made up in the hot water generated by cooling a previously sterilized batch of medium. Apart from the obvious saving in energy, reducing the time to reach sterilization temperature reduces the heat-damage to the medium. When the correct amount of hot medium has been prepared, the fermenter is closed and the temperature is raised. The heating is accomplished by:

1   Injecting steam into the medium, therefore the medium is prepared slightly concentrated to compensate for dilution by the condensed steam, and boiler additives have to be carefully monitored to ensure no adverse effect on the growth of micro-organisms. Heating by steam injection is very effective, by releasing its latent heat of vaporization as it condenses in the medium. Steam is supplied both through the air line and the sample port.

2   The medium is also heated by conduction by passing steam through the attemperation jacket (and coil, if fitted).

The stirrer must be operated throughout the entire sequence of heating, sterilization and cooling, to improve heat transfer, and to prevent medium being baked onto the hot surfaces.

When the medium reaches 100°C, indicated both by a temperature gauge and by vigorous evolution of steam from the exhaust, the exhaust valve is closed to allow the pressure and temperature to rise to the sterilization level; 120–150°C depending on various factors, including the size of the vessel and heat-stability of the medium. The sterilization process so far has been essentially the same as for a laboratory autoclave, but for cooling a fermenter, it is essential to avoid the formation of a vacuum within the vessel. It is possible, particularly in the case of an empty fermenter, that a vessel under vacuum could be deformed by the external air pressure, but the main reason for avoiding pressure below atmospheric within the fermenter is that the nosterile air of the environment could be drawn through any small leak in the welded seams of the vessel or pipework, or through joints or valves. Therefore it is essential to admit a supply of sterile air to the head-space above the medium as soon as the sterilizing steam supply is stopped. For the same reason, the cooling water supply through the jacket or coil must be sterile (by chlorination, UV irradiation or filtration) to avoid contamination from chance leakage into the vessel. Admittedly there should be no such leakage, but it is good economic sense to avoid loss of the expensive sterile medium in a fermenter should it occur. Cooling, inevitably, is a slow process, entirely by heat transfer between the medium and the cooling coil or jacket; the heated water is used for preparation of a following batch of medium. Finally the inoculum is added and aeration and attemperation are adjusted to suit the cultural requirements of the organism.

There are advantages to this system, which reduces the time the fermentation vessel is unproductive: an empty fermenter is sterilized relatively quickly. Expensively high standards are required of a fermenter, which has to operate aseptically for up to a week without any possibility of maintenance. One simpler, cheaper cooker vessel can provide the sterile medium for a number of fermenters in succession. On the other hand, a fault in the cooker puts all of its fermenters out of action, and there may be long sterile transfer lines to the more distant fermenters.

The clean fermenter is supplied with steam through a valve at the top and air and condensate are drained from a valve at the lowest point of the base of the fermenter. Often the bottom valve is fitted with a thermostatic steam trap: when air and water are exhausted, the higher temperature of pure steam closes the valve, then the pressure is raised to hold the temperature at 120–150°C for the required time. The vessel may be cooled under pressure of sterile air by passing cold water through the coil or jacket, but alternatively cooled sterile medium may be added immediately after sterilization of the vessel.

The medium may be sterilized (1) in a batch cooker of the same size as the fermenter or (2) by a continuous high temperature short-time process through a heat exchanger.

1 Sterilization of medium in a cooker of the same size as the fermenter: a closed vessel, fitted with a coil for heating and cooling and an agitator to maximize heat exchange, is filled with medium and sterilized as described above. The pipework joining the cooker and fermenter is sterilized by introducing steam at each of the high points of the system, and drained of air and condensate at each low point. The three sections, cooker, pipeline and fermenter, are separated by valves so that each can be sterilized independently if necessary (Fig. 6.4a).

When the three sections have been sterilized, but not necessarily cooled, the sterile medium is transferred to the fermenter by a combination of gravity and pressure of sterile air. The cooker is at a higher level than the fermenters it services, and the sterile air pressure in the cooker can be raised to, for example, 2 bar in the cooker and 0.3 bar (above atmospheric) in the fermenter. So the transfer is accomplished without pumping sterile medium, an operation to be avoided if possible. In some establishments the connecting pipeline is disconnected after the transfer, to avoid any accidental contamination of the fermenter when the cooker is re-charged with nonsterile medium for the next fermenter in the sequence. Alternatively, a pipeline left permanently in position is held under steam pressure to sterilize any such leaks.

By this system, the medium can be sterilized in the cooker

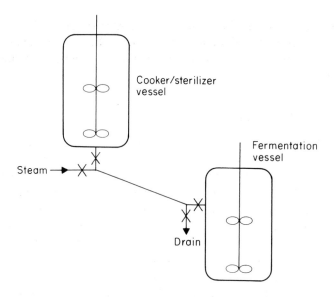

Fig. 6.4.(a) Batch sterilization of medium in cooker vessel.

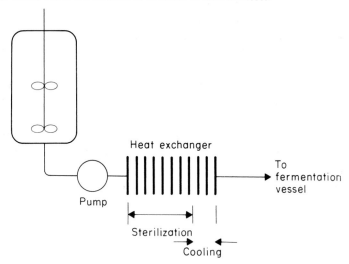

Fig. 6.4.(b) Continuous sterilization of medium.

before the previous fermentation has been harvested, and the fermenter can be emptied, cleaned and resterilized while the sterile medium is cooling. The system reduces to a minimum the time each fermenter is out of service, an important consideration with such expensive units.

2   Continuous sterilization. The medium, prepared nonsterile in a mixer vessel, is passed through a heat exchanger at such a flow rate that it is in contact with the heated portion for sufficient time to sterilize. Note that the pump required for transfer of the medium (Fig. 6.4b) is on the nonsterile side of the heat exchanger: equipment requiring regular maintenance must not be fitted unneces-

sarily to a sterile system. Continuous sterilization requires a longer time than the transfer of already sterilized medium from cooker to sterile empty fermenter, but causes less heat damage to the medium.

Inoculation of fermenters

In most fermentation processes the culture organism is grown specially as inoculum for each fermentation, and discarded at the end. Brewing is exceptional in this respect, in that part of the yeast is re-used as inoculum of the next fermentation. In other fermentations, genetic instability of the culture, the risk of building up contamination over successive fermentations, or the low viability of cultures at the end of fermentation, prevent such practice.

Stock cultures are kept in the laboratory in freeze-dried form, deep frozen in liquid nitrogen, on silica gel pellets, or in some other suitable system to maintain viability and the industrially useful properties. The initial inoculum is added to a relatively small volume of medium, for example 1 litre in a 2-litre Büchner flask. The culture is grown under optimal conditions for growth, which need not be the same as for development of product, and will almost certainly be over a shorter incubation time. A small sample of the 1-litre culture is used to check that it meets industrial specification; the remainder is used to inoculate a 20-litre fermenter (Fig. 6.5).

Cultures are then grown in successively 20-fold larger amounts to reach the scale of the inoculum of the final production fermenter. The successive inoculations are by the same principle as transfer of sterile medium to a fermenter (Fig. 6.4a).

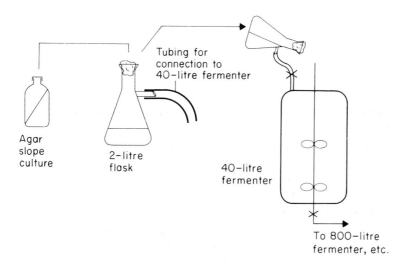

Fig. 6.5. Inoculation sequence.

Customarily the inoculation level for the final stage is high, often 1/10 and in some plants as much as 1/5 of the fermenter volume. This is for two principal reasons. If any contamination of the production medium has occurred (and, being the greatest volume in the series, this is where such an accident is most likely) the large inoculum has a better chance of outgrowing and suppressing contaminants. More importantly, with a large inoculum the time taken to complete the fermentation is reduced. The total amount of microbial mass produced depends on the properties of the medium, not on the size of inoculum, but with a larger inoculum, maximum growth is achieved more rapidly (Fig. 6.6). Antibiotic fermentations are continued through a period of maturation of the culture, but the final maximum yield of antibiotic is achieved correspondingly earlier in culture 1 of Fig. 6.6. Therefore the fermenter is harvested and available for the next fermentation more quickly. Given the high capital and operating costs of fermentation equipment, this is obviously good economics.

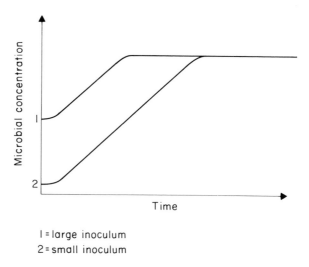

1 = large inoculum
2 = small inoculum

Fig. 6.6. Effect of large (1) and small (2) inoculum on reaching the total final cell concentration.

Accessory equipment

Various accessory items are required; the following list is not exhaustive but is intended to illustrate the principles involved in the design of fermentation equipment.

1   *Safety valve.* Since the fermenter is operated under pressure it requires a safety valve. The standard spring-loaded valve as fitted to a boiler is a hazard to a sterile system: micro-organisms can pass between the open valve and its seat. Such a design is permissible on steam lines, or unsterile air lines, but sterile or

pure culture zones are protected by a metal foil disc (Fig. 6.7a) constructed to burst at the maximum allowed pressure of the fermenter, and installed as a last-resort protection against explosion. Normal operation maintains pressures well below the burst-pressure of the disc.

2 *Pressure gauge.* A similar principle is applied to pressure gauges fitted to sterile or pure-culture systems (Fig. 6.7b). The standard design of pressure gauge operates by uncoiling of a capillary tube according to pressure, but the dead-end of the tube cannot be sterilized. Therefore the diaphragm gauge is necessary: distortion of the metal foil or heat-resistant synthetic rubber diaphragm operates the gauge, but there is no risk of contamination provided the diaphragm is renewed at the correct service intervals.

3 *Control valves.* Of the various types of valve used to control flow of fluids, most are prone to leakage and are therefore unsuited to pure-culture work. Although not the only suitable design, one of the best, and most frequently used, is the diaphragm valve (Fig. 6.7c). In the closed position, the synthetic rubber diaphragm is

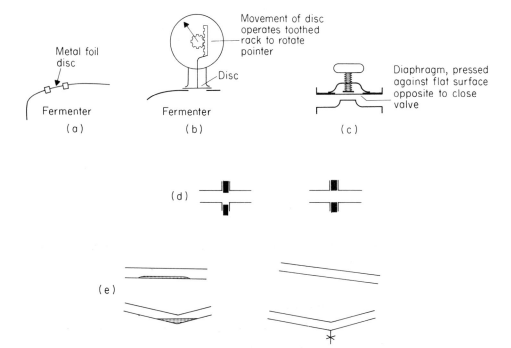

Fig. 6.7.(a) Safety valve for sterile system.
b) Pressure gauge for sterile system.
(c) Diaphragm valve for sterile system.
(d) Bad (left) and good (right) alignment of pipework joints and gaskets. The badly finished joint allows microbial soil to collect in the depression.
(e) Horizontal pipework or dips without drainage collect spent culture or other microbial soil. Correct design shown on right.

pressed against the flat inner face of the valve. Provided the diaphragm is undamaged, there is no passage between the contents of the valve and the nonsterile environment. Diaphragm valves are convenient for both manual and automatic operation. Automatic control on large fermenters is normally an electropneumatic system: the (weak) electrical signal controls a supply of sterile air (which is always available, for aeration purposes, in a fermentation plant) to provide the pneumatic power for operating the valve.

4   *Pipework.* Careful design of pipework reduces the possibility of microbial contamination. Pipework should normally be welded, with smooth joints, but at points where periodical dismantling is necessary (for example at pumps or valves which are removed for maintenance) bolted joints are used. Such joints have to be accurately aligned to avoid any internal irregularities which cause an accumulation of microbial soil (Fig. 6.7d). In addition, pipework should if possible have a continuous slope in one direction: horizontal sections or dips collect spent medium or washings, a source of infection (Fig. 6.7e). Any low point must be fitted with a valve to drain off washings or condensate.

5   *Temperature control.* Resistance thermometers, which measure the change in resistance of a platinum coil with temperature, have ideal properties for fermentation equipment. They can be calibrated for greatest accuracy over the normal operating temperature of the fermenter, usually around 30°C; also, any fault is recorded as low temperature, preventing accidental overheating of the culture. The probe is installed in an oil-filled well on the side of the fermenter; it does not have to be in direct contact with the culture. The signal from the probe can be read directly (on an ammeter calibrated to read °C) and can operate recorder or control equipment.

6   *pH control.* pH electrodes must, obviously, be in contact with the medium and cannot be removed or replaced during a fermentation run. Modern steam-sterilizable electrodes are sufficiently reliable for such duty, but for safety, often a duplicate replacement electrode is installed. Most fermentations involve acid production during growth and adjustment of pH is by automatic addition of sterile alkali (or, in some installations, addition of NH_4OH as both N source and neutralizing agent) under the control of the pH recorder–controller unit.

7   *Foam control.* Ingredients of microbial culture media, especially polypeptides and proteins, are surface active compounds and the vigorous aeration and agitation necessary for efficient growth generates copious foam. Without foam control, the entire contents of a fermenter could be lost through the exhaust. Modern antifoams are often silicone compounds, surface active agents in their own right, but which are unable to produce stable foams. Therefore, by competitively replacing the foam-forming

medium components or microbial products, foaming is
suppressed.

There are two approaches to the suppression of foam. Addition
of antifoam to suppress existing foam is apparently the simplest
method, and can be automated by a simple relay system (Fig. 6.8).

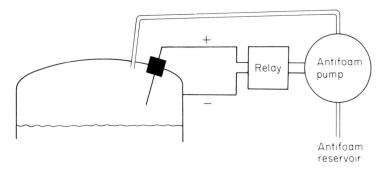

Fig. 6.8. Insulated probe system for detection and control of foam.

Foam completes the circuit between insulated probe and the
vessel, starting a pump supplying sterile antifoam solution. Since
quenching of foam is slow, an interrupter device is normally fitted
to add antifoam only 5 sec of every minute the circuit is complete.
The disadvantage of the system is that even with interrupted
addition, the amount of antifoam required to quench foam already
formed is much greater than the amount to prevent foaming.
Modern computer controlled equipment, programmed from experience with pilot-scale fermentations, adds just sufficient
antifoam at the correct times to prevent serious foaming. By its
surface-active effect, antifoam seriously reduces oxygen transfer
rates and excess addition cannot be tolerated.

8 *Sampling points.* In all fermentations it is important to take
samples, aseptically, as necessary to monitor the process. Figure
6.9 shows two designs of sample point commonly fitted to fer-

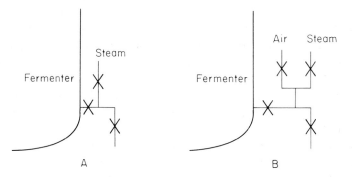

Fig. 6.9. Alternative designs of sampling point. A requires a continuous steam
supply for protection against contamination. B is normally protected by sterile
air under pressure and steam is required only for sterilization after sampling.

menters. *A* is suitable for large installations where steam is continuously available. In smaller installations the steam supply is available only for sterilization, therefore system *B* is protected from external contamination by sterile compressed air. The sequence of sampling is to turn off the supply of steam (in *A*) or air, (in *B*), and flame the sample nozzle. After running a small volume to waste (which cools the system and removes stagnant medium in or near the sample point) the sample is collected in a sterile container. The remaining liquid in the pipe is blown out by steam which then resterilizes the sample point under pressure.

Air-lift fermenters

Two disadvantages of the stirred-vessel system are the high energy requirement for agitation and the damaging effect on cells of the shear effect. The air-lift fermenter overcomes both difficulties by creating a rapid, low-shear movement by the effect of aeration. The liquid movement is initiated by injection of air at the foot of the riser column (Fig. 6.10[A]). When the liquid is in rapid movement the air valve is opened on the down flow column. Air is trapped in the downward flow of liquid and solution is enhanced by compression. In the riser column the compressed air bubbles expand, as they rise, maintaining the rapid movement of the liquid after the start-up air has been turned off. For most effective circulation of air the tube should be tall, at least 10 m, which corresponds to 1 bar differential between top and bottom. Because of the rapid movement over the jacket cooling system heat transfer is very efficient and internal cooling coils are unnecessary. The same principle is achieved in more compact form with the two columns arranged concentrically (Fig. 6.10[B]).

The air lift principle has been particularly useful in the process for production of single-cell protein from methanol (the 'Pruteen' process of ICI); aeration and attemperation requirements would have been impracticable in mechanically stirred fermentation vessels of the scale used (> 1000 m^3). The principle has also been applied on a smaller scale to the culture of plant and animal cells, to avoid the damaging shear effects of the stirring gear.

Although not strictly an air-lift fermenter, stirred vessels of the Waldhof design incorporate an inner draught tube to generate a continuous vertical cycling, and by drawing foam back into the medium, to reduce foaming.

Continuous fermentation

The majority of industrial fermentations are operated as batch processes: the medium is inoculated, the culture grown, the product harvested and the cells are discarded at the end of the

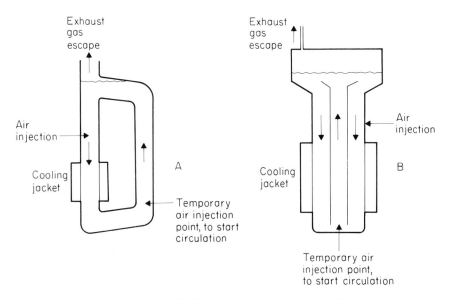

Fig. 6.10. Alternative designs of air-lift fermenter; (A) with separate upflow and downflow; (B) with concentric system.

process. For various reasons, including the fundamental inefficiency of providing nutrients to grow cells for each batch, continuous fermentation processes seemed to offer economic advantages over batch processes. The wide interest in laboratory-scale continuous culture in the 1950s stimulated various attempts to apply the technique on a larger scale.

The basic principle of continuous culture is applicable to any scale of operation, especially in the form of the chemostat, which has been the most successful type on the laboratory scale (Fig. 6.11a). Sterile medium is supplied at constant flow rate to the homogeneously stirred culture vessel, and culture overflows at the same rate for harvesting. Except where the organisms themselves are the desired product, the simple 'open' system with free escape of cells may be modified for more economical use of nutrients as a 'closed' system whereby only culture medium escapes freely; cells are retained. Often this is achieved by recovery of cells by settling or continuous centrifugation and returning cells to the vessel. In practice 100% return of cells is impracticable, since cell concentration is raised to a level preventing further cell growth, but partial recovery and return of cells is compatible with longterm continuous operation.

Large scale continuous fermentations have been largely unsuccessful for a number of reasons. Attempts in the brewing industry from 1958 onwards were technically successful, but the continuous process has subsequently been abandoned in most countries. The fermentation product is a complex mixture of ethanol and numerous minor but important metabolic byproducts which con-

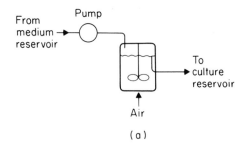

(a)

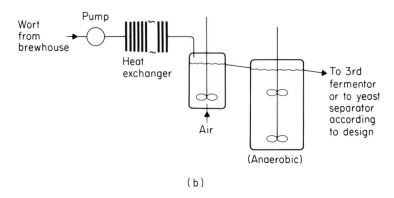

(b)

Fig. 6.11.(a) Simple single vessel continuous culture system ('chemostat').
(b) Multivessel continuous culture system as applied to brewery fermentation.

tribute to flavour, and the different conditions of continuous fermentation gave beer of different flavour from the 'traditional' batch product. One vessel was insufficient; at least two, and often three, were required to separate the initial, aerobic, stage of yeast growth from the anaerobic fermentation (Fig. 6.11b). Many beers are produced by a mixture of yeast strains, not by a single yeast, and the mixed cultures could not be maintained in correct ratio over longterm continuous culture. Even with a single pure culture, longterm changes in the properties of the yeast, as faster-growing mutants swamped the fermentation, could cause problems. Finally the introduction of the cylindroconical fermenter in the brewing industry gave the advantage of fast batch fermentation without the complexity of the expensive stirred fermenters required for the continuous process. Even the tower fermenter (Fig. 6.12), with fewer moving parts, could not be justified. In the tower system a single vessel sufficed, since the aerobic stage at the base of the fermenter was separated in the nonhomogeneous conditions from the aerobic stages at higher levels. However, the commitment to one single yeast strain throughout a continuous run was a serious disadvantage compared with the versatility of batch fermentation.

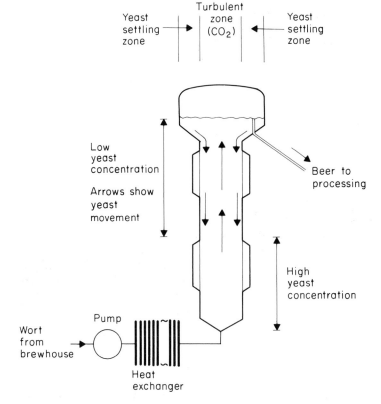

Fig. 6.12. 'Tower' fermenter applied to continuous brewery fermentation.

In the antibiotic industry, continuous fermentation was rejected because of the genetic instability of cultures which have been extensively mutated and selected for high yields. The development of batch inoculum is planned to involve the smallest possible number of generations, since each nuclear division represents a risk of culture degeneration. Such degeneration would be inevitable over a long period of continuous culture. Furthermore, antibiotic production is normally associated with secondary metabolism late in the course of batch growth cycle. Therefore multivessel installations would be required, with a low flow rate of culture medium, and could not have been economical.

The best potential for the use of continuous culture is for processes in which the organisms are themselves the required product. This principle does not apply to bakers yeast, which must meet enzymic specifications easily lost on continuous culture, but has been applied successfully to production of micro-organisms as single cell protein. Another potentially useful application, although not fully exploited to date, is to products associated with the logarithmic growth phase of the organism. In continuous culture the organisms are maintained indefinitely in late log phase growth. Finally, despite the complications of continuous fermentation of alcoholic drinks, continuous production of industrial, for example fuel, alcohol promises to be effective with suitable substrates, for example molasses. Since the subsequent distillation process is continuous, this is a further advantage of continuous fermentation.

Immobilized cell reactors

Numerous applications have been developed of the use of immobilized enzymes. The principle of immobilizing whole cells in a column has the advantage of avoiding the preliminary extraction of the necessary enzyme, or where more than one enzyme is involved in the reaction, all are held correctly together within the cells. The 'closed' system of continuous fermentation is, in effect, an immobilized cell reactor since cells are prevented from leaving the system. Experience showed a limited operation life, maximum 10 days, to such nongrowing systems. A longer activity of cells is achieved by forming them into beads with alginate gel, or by holding the cells within the interstices of porous material, where a limited growth is possible without damaging the support.

The obvious application to alcohol fermentations has suffered from the flavour problems discussed in the context of continuous fermentation, and from the destructive effect of bubbles of CO_2 when the process is attempted on a large scale. Steroid conversions have been successfully operated with immobilized cells; an important advantage of the immobilized system is that steroids, poorly soluble in water, can be applied at higher concentrations in solution in organic solvents. The vinegar generator and the trickle filter for effluent treatment are also examples of immobilized cell systems, both with a long history of successful operation. Otherwise, successful and economical large-scale operation has yet to be achieved with immobilized cell systems. Antibiotics, for example, are not produced by such a method because of the complexity of their production. Cysteine and valine form the nucleus of the penicillin molecule but cannot be condensed to penicillin in an immobilized system. On the other hand, there are many types of process, at present practicable on a laboratory scale, which could be predicted to operate commercially with immobilized cells in the future.

Chapter 7. Downstream processing

After fermentation, the product must be recovered and purified from the fermentation broth. The methods are obviously dependent on the nature of the product: volatility, heat-stability, concentration and location (cellbound or extracellular) are important considerations. Methods commonly used in the fermentation and other biotechnological industries are distillation, centrifugation, filtration, solvent extraction and various forms of chromatography.

Distillation

This method has a long history, in its association with distilled spirits; production of fuel alcohol, acetone and acetic acid (for distilled vinegar) are other examples. The production of potable spirits illustrates both common types of distillation; a batch process ('pot stills', in the jargon of the Scotch malt whisky industry) and a continuous process, invented by Coffey in 1831, the Coffey still.

The batch distillation process requires two (or, rarely, three) stills in series but is not an efficient process in terms of strength or purity of the ethanol produced. In the production of quality whiskies, brandies, rums, etc. the chemical inefficiency is a desirable feature of the process: not only ethanol, but traces of other volatile organic compounds, each contributing to the final flavour and aroma of the product, are collected in the distillate, and the 65–70% ethanol content is suitable for the subsequent maturation without further dilution. Figure 7.1, of typical malt whisky pot stills, shows the condenser heads inclined downwards, but they could equally correctly be drawn horizontal or inclined upwards. The geometry of the unit in this respect, and in the shape of the neck of the still (bulbous in some installations) influences the extent of codistillation of 'congener' organic compounds with ethanol and so the character of the product. The names used in annotating Fig. 7.1 are those commonly used in Scotland.

'Grain spirit' produced by fermentation of a mash of malt and unmalted cereal is obtained by continuous distillation, as is the spirit for gin, vodka and various liqueurs which require a base spirit of low flavour. In the production of industrial-quality ethanol, for example as fuel, continuous distillation produces a

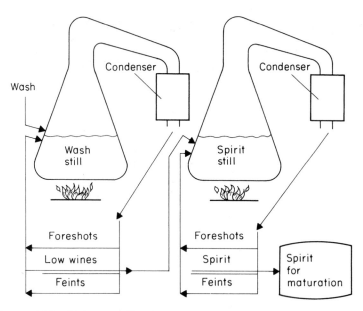

Fig. 7.1. Malt whisky pot stills. Three fractions are collected from each still; fractions 1 and 3 are returned for redistillation with the next batch and only the middle fraction continues to the following stage.

satisfactory product as economically as possible but 96% (v/v) ethanol, the azeotropic mixture is the highest concentration possible by distillation of aqueous ethanol. Normally distillate of 65–70% ethanol is collected for production of spirit drinks. Anhydrous ethanol, required for many industrial purposes and certainly for blending into petroleum spirit as motor fuel, is produced by further distillation from benzine.

The Coffey still (Fig. 7.2) is a cyclic system requiring two columns, the 'rectifier' and 'analyser', each a stack of 30–32 perforated plates. The wash from the fermentation is heated, during its passage down the rectifier column, by the upflow of hot vapour. The hot wash is discharged into a trough at the top of the analyser column, and the liquid, falling down that column, is heated by the upflowing steam. The resulting hot vapour is piped to the base of the rectifier and condenses on the coils carrying the incoming wash. In the temperature gradient up the height of the rectifier, each organic compound condenses at its appropriate height, corresponding to the temperature at that level. The plates of the column are perforated to allow free ascent of the spirit vapour, but any liquid ethanol falling below the collection level is vaporized and returned to the outflow.

The principle of the Coffey still can be applied to other fermentation products, provided they are volatile heat-stable compounds of lower boiling point than water. Alternatively, a volatile deriva-

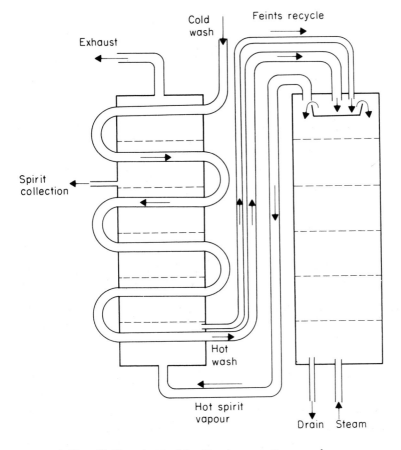

Fig. 7.2. Coffey still. Note that 6 of the 30 column sections are shown.

tive may be purified by distillation. This principle is used in the production of food-grade lactic acid by preparation of methyl or ethyl lactate, distillation, then hydrolysis of the pure ester.

Flotation

This method is largely of historical interest and the only important present application is the recovery of yeast from traditional beer fermentation. In the course of the fermentation, cells with attached bubbles of CO_2 rise to the surface as a firm yeast head. Part of the yeast population sinks back into the fermenting beer, but the excess yeast is skimmed off on at least two occasions during the fermentation to be re-used as inoculum for a following fermentation. Until the development of reliable filtration and centrifugation equipment in the late 19th century, bakers' yeast was also recovered by this method.

Precipitation and flocculation

Uniform suspensions of cells may be caused to flocculate to aid separation from the medium; in the production of beer, wine and cider it is common practice to use yeast strains which grow in even suspension during the active fermentation but spontaneously aggregate, i.e. flocculate, at the end of the fermentation to leave a clear product. Fining agents may also be employed: the best known is isinglass, a collagen-type protein derived from the swim bladders of large tropical fish. Isinglass, of opposite electric charge to yeast cells, precipitates the cells to a firm deposit.

Citric and lactic acids produced by fermentation are precipitated as their calcium salts, by addition of lime or chalk to the medium after fermentation. The precipitate, usually collected by filtration, is treated with sulphuric acid to recover the acid.

Protein (including enzyme) or polysaccharide products are also conveniently and economically concentrated and purified by precipitation: typically ammonium sulphate or organic precipitants (ethanol, acetone) are used. The procedure is essentially a large-scale version of preparative laboratory methods.

The extraction and purification of riboflavin from fermentation broths shows an interesting variation on precipitation. Riboflavin is soluble in water but the reduced form, obtained by addition of stannous chloride solution, is insoluble. The harvested precipitate is washed and then re-oxidized by bubbling air through the precipitate to redissolve.

Precipitation and flocculation methods have the advantage of requiring no energy but are slower than mechanical separation. Also, in many processes leading to food or health care products there may be no possible (safe) precipitants or flocculants available. However, the principle is widely applied in effluent-treatment, where many different chemical precipitants or flocculants are used.

Filtration

Although modern designs of filter have a history of barely 100 years, filtration has traditionally been an essential step in the production of brewer's wort. Here, the husk material of the ground barley malt, or the spent hops, are supported on the perforated bottom plates of the strainers, to filter out particulate suspensions and leave a clear wort for fermentation. Until recent times, however, postfermentation filtration of beer or other alcoholic beverages was unknown and clarity was achieved by choice of suitable flocculent yeast strains and the use of isinglass or other fining agents.

Plate-frame filter

A common type of filter in the beverage industries is the plate frame filter (Fig. 7.3); the primary filter material is a heavy cloth of cellulose material or asbestos (use of the latter having recently been discontinued even though the asbestos used was not the blue carcinogenic type).

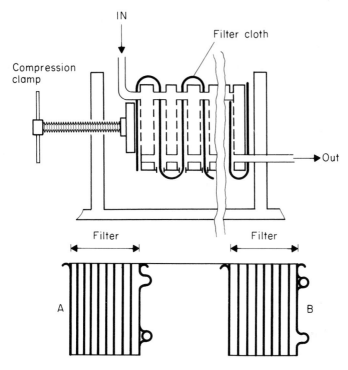

Fig. 7.3. Plate frame filter. Alternate plates of patterns A and B are interleaved with continuous filter cloth to provide a large surface area.

The plate-frame filter is a typical 'depth filter', micro-organisms and other particulate material being trapped in the passages through the cloth by electrostatic effects. The effective pore size is larger than the diameter of the particles or micro-organisms. It is customary for the feed to plate frame filters to be dosed with a slow bleed of a diatomite slurry as a filter aid. Diatomite rock, fossilized silica shells of diatoms, ground to uniform particle size, functions as a subsidiary filter, trapping micro-organisms and inert particles in a filter bed that gradually increases in thickness with time. Ultimately the inlet side of the filter becomes blocked with diatomite, bringing that filter run to an end. The main advantage of diatomite filter aid during a long filter run is that, by trapping particles before reaching the filter cloth, 'blinding', i.e.

blockage of the pores, is prevented. Since the preparation of the filter before a run, and cleaning afterwards, are labour-intensive operations it is important that the filter be operative for long periods. Therefore it is customarily used after a preliminary clarification, by settling or coarse prefiltration, and operated at 0–2°C to prevent microbial growth in the filter. As normally operated, the plate filter does not guarantee sterility, but the filtrate is of very low microbial count, normally < 10 cells per litre.

The main application of such filters is in the production of alcoholic and clear soft drinks; in both types the liquid is already of low solids content. The filter-aid is difficult to dispose of after filtration, and although in some areas it is collected for incineration and re-use, normal disposal is by burial as landfill. The filter cloths are normally cleaned and re-used repeatedly.

A variation on the plate-frame filter uses not filter cloth but a nest of perforated stainless steel plates, or a stack of closely-spaced stainless steel rings. These supports do not themselves provide any filtration, but support the diatomite. A slurry of diatomite in water is first applied to build a filter bed, then filtration proceeds as previously described. Such filters, for example the Metafilter, are not as expensive in labour costs as the cloth filter, but are still sufficiently labour-intensive to be limited to the same applications. Their main advantages are more robust construction, and easier and cheaper cleaning and sterilization.

Rotary vacuum filter

Plate filters are operated continuously, but only for a limited period until full of filter aid. In contrast, rotary vacuum filters have, theoretically, the capability of indefinitely long operation, but in practice are limited by the risk of development of bacterial infections.

The simplest system is a drum with a hollow spindle connected to a vacuum pump, and hollow spokes evacuating through the perforated metal circumference of the drum. The filter medium is commonly woven cloth, of similar texture to that of the plate filter, but closely-spaced metal wires or coils are used in some designs to assist separation of dried material. Clarification of penicillin broth is achieved by running the culture, with filter aid added if necessary, into the trough: medium is aspirated into the spindle and so to processing, while solids collect on the circumference to be scraped off as shown in Fig. 7.4a. The alternative arrangement in Fig. 7.4b dislodges the dried mat without the use of a knife.

The special requirements of the bakers yeast industry prevent use of diatomite filter aid, but the drum is precoated with food-grade starch which performs a similar function.

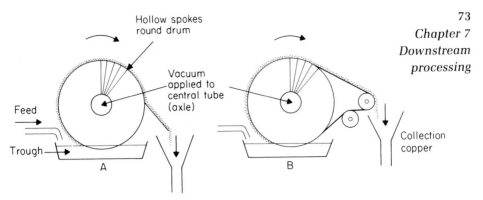

Fig. 7.4. Rotary vacuum filters, showing alternative systems of removal of microbial mat. (A) scraper blade; (B) flexing of porous belt.

Centrifugation

Centrifugation is of limited value in postfermentation processing, largely because of the energy and maintenance costs compared with precipitation or filtration. Nevertheless, centrifugation may be the only practicable separation of cells from viscous media, or when filter-aids cannot be used. Continuous centrifugation with intermittent automatic discharge of solids from the rotors is now the commonest system, being capable of prolonged continuous operation. A typical example is illustrated in Fig. 7.5; the solids are discharged intermittently into an annular receiver by briefly opening the spinning rotor. In the design shown, the solids are deposited on the surfaces of the stack of internal cones and are forced radially outwards for collection at the extreme circumference of the bowl.

Normally centrifugation is only a postfermentation process, and although operated under clean conditions, aseptic operation is unnecessary. Recycling of part of the cell population in continuous culture employed continuous centrifugation in some versions of the system, unfortunately increasing the difficulty of maintaining pure culture conditions throughout the period of the fermentation.

Extraction methods

Various processes are available for the specific extraction of product from fermentation broths, or other sources of natural products. The commonest and most useful methods are solvent extraction and a variety of chromatographic methods: absorption chromatography, ion-exchange chromatography, high-pressure liquid chromatography, gel filtration, affinity chromatography. The most suitable system for any one product, and the best con-

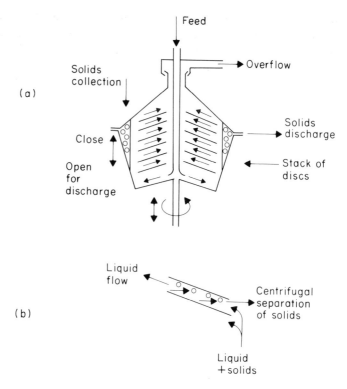

Fig. 7.5. Continuous centrifugation.
(a) Self-emptying centrifuge.
(b) Pattern of separation of solids on angled stack of rotating discs.

ditions under which to use that system, are determined in the course of the extensive development work before commercial production, often by trial-and-error. The following examples are presented to show typical applications of the various methods available.

Solvent extraction

Many extraction processes depend on selective solution from the fermentation broth by a water-immiscible solvent. In the extraction and concentration of penicillin, described here as an important example, the favourable partition coefficient between organic and aqueous phases is achieved by alterations of pH. In the case of steroid extractions, such pH adjustment is unnecessary.

The clarified penicillin broth, after removal of mycelium and ammonium sulphate-precipitable material, is cooled to 0–2°C and fed into a continuous-flow counter-current solvent extraction unit. Immediately before solvent extraction the broth is acidified to pH 2.5 with sulphuric or phosphoric acid to permit extraction into butyl acetate or alternative solvent. The solvent:broth ratio is approx. 1:10 to provide a tenfold concentration. At pH 2.5, virtually 100% extraction into the organic phase is achieved rapidly (Fig. 7.6) and, since penicillin is very unstable at that low pH, the butyl acetate is quickly separated and neutralized with one-tenth of its volume of the calculated strength of NaOH solution, which concentrates a further tenfold. The solution of the sodium salt of penicillin is crystallized to purify the product further. Meanwhile the solvent is returned to repeat the cycle through the extraction system.

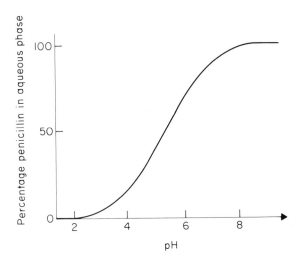

Fig. 7.6. Partition of penicillin between water and butylacetate according to pH.

Absorption chromatography

Most laboratory methods of chromatography can be scaled-up for extraction of fermentation products, and are especially valuable for high-value products at relatively low concentrations in the culture broth. Absorption chromatography is the simplest system, used in the laboratory as paper chromatography. On an industrial scale, columns of activated carbon, alumina, silica gel, apatite or various nonionic synthetic resins constitute the solid phase. An early example of column chromatography was the absorption of streptomycin from clarified fermentation broth on to activated carbon. The product was then specifically eluted by methanolic HCl.

A recent development in preparative separation is the application of high-performance (or high-pressure) liquid chromatography (HPLC) but at present it is an expensive technique applicable only to very high value (over £1000/g) products in kg amounts, for example minor blood components.

Ion-exchange chromatography

This is a more predictable extraction and concentration method than absorption chromatography, provided the required product is capable of reacting with an appropriate ion-exchange resin. Many types of anionic and cationic resins are now available, although for most applications only one type or the other would be predicted to be suitable. Even so, a range of resins and conditions for absorption and elution have to be investigated to find the most efficient and economical process.

Ion-exchange has now replaced absorption on carbon in the processing of streptomycin. Clarified broth is passed through a cationic resin in the sodium form; streptomycin is absorbed and Na is displaced as NaOH. When the resin column is fully saturated, the flow of broth is directed to the next column of the series while the first is treated with the minimum volume of dilute acid (for example H_2SO_4 to yield streptomycin sulphate). Subsequently, the resulting acid form of the resin is regenerated by NaOH solution. This principle has also proved suitable, although using different absorption and elution conditions, for other antibiotics and various vitamins and nucleotides.

Gel filtration and affinity chromatography

These are useful laboratory methods which have not so far been applied to large scale production. However, there are many examples of low-volume high-value products which justify the use of these expensive and labour-intensive methods. Gel-filtration, allowing the selection of an eluate fraction containing molecules of a specific molecular size, is applied to the preparation of blood products. Affinity chromatography, to a column treated with a highly specific absorbent for the product (for example enzyme for substrate or vice versa; antigen for antibody or vice versa). Proteins, enzymes, immunological reagents and blood products may all be obtained in commercial quantity, in a highly purified state, by affinity chromatography. For example immunoaffinity chromatography has been used to purify a number of therapeutically important proteins including interleukin 2, tissue plasimogen activator and interferon. The efficiency of this system is well illustrated by its application to the purification of interferon where a single pass down an anti-inferon column resulted in a 5000 times purification of this protein from cell culture supernatants.

Some recent advances

The types of matrices available for column chromatography now include cellulose, agarose, acrylamide, dextran and copolymers of these and a number of inorganic materials of which the most popular are silica based systems. The availability of extensively cross-linked polymers has led to the production of beads which are less deformable, and therefore allow higher rates of flow, are more resistant to attack by alkali, which improves the sterilization properties and yet retain their hydrophilic characteristics which are of significance in protein separation. The range of chromatographic purification methods available is steadily increasing and now includes hydrophobic, inverse hydrophobic, metal chelate and affinity systems employing lectins and antibodies in addition to the more traditional gel filtration and ion exchange systems. In time some of these may prove suitable for production purposes.

Purification protocols have also benefited from the attentions of the geneticists. As mentioned earlier the addition of a signal peptide leader sequence to a cloned gene allows excretion of the protein concerned into the periplasmic space of *E. coli* with associated benefits in protection from proteolytic degradation and relative ease of purification. The addition to the C-terminus of a cloned protein of a 'tail' of arginine residues renders the whole protein more basic and allows primary separation to be carried out on a cationic ion exchange matrix.

In *E. coli*, high levels of expression of eukaryotic genes (including urogastone, interleukin 2, prochymosin, interferons and animal growth hormones) often lead to the accumulation of cytoplasmic inclusion bodies. These are granules of insoluble proteins which can be sedimented by centrifugation after cell disruption and then washed to provide a relatively pure source of material for further purification. These must be solubilized and renatured before further treatment is possible and the procedures adopted tend to be specific to a particular protein. Solubilization often requires treatment with alkali, urea or guanidine hydrochloride.

Chapter 8. Enzyme technology

There are two broad aspects to the commercial application of microbial enzymes. The original enzyme industry centred on high volume, low-cost enzymes for use in the food-processing and related industries. A later, significant development was the incorporation of proteases in household washing detergents. More recently, however, high-cost, purified enzymes for use, particularly as diagnostic reagents in clinical biochemistry are being introduced into the lucrative health care market.

Industrial enzymes

The bulk industrial enzymes are prepared almost exclusively from Gram-positive bacteria and fungi (Table 8.1). They are generally extracellular enzymes which are secreted into the medium in massive amounts usually by strains selected from traditional mutation programmes. They are purified by precipitation and filtration to remove undesirable metabolites and particulate matter and to improve stability and standardize activity but often no attempt is made to purify them beyond this. Indeed these products are usually enzyme mixtures and for most applications this will be desirable. The market is dominated by starch hydrolysing enzymes and proteases which together account for over 80% of sales.

Immobilization of enzymes

Enzymes are often used as nonrecoverable chemical reagents, in which case they are added to the substrate, incubated at the required temperature and pH for a period and subsequently destroyed. Amylases, proteases and other inexpensive bulk enzymes are used in this way. Alternatively, enzymes may be attached to an inert support (immobilized). This offers the advantages of (1) recovery and re-use of the enzymes, in batch reactors; or (2) the development of continuously operated enzyme reactors similar to continuous fermentation systems used for micro-organisms; (3) the possibility of multi-enzyme systems; and (4) the enzyme does not remain in the processed solution. However, there are some disadvantages, the enzyme may be stabilized by immobilization but it may also lose activity, and the process becomes technically more complex.

Table 8.1. Some common industrial enzymes, their sources and uses.

Enzyme	Source	Principal uses
α-Amylase	*Aspergillus oryzae*	Starch hydrolysis for sugar
	Bacillus amyloliquefaciens	syrups, brewing
	Bacillus licheniformis	Textiles and paper
Cellulase	*Aspergillus* sp.	Fruit and vegetable
	Trichoderma rusei	processing
	Penicillium sp.	
β-Glucanase	*Bacillus subtilis*	β-Glucan hydrolysis in
	Aspergillus niger	brewing
Glucoamylase	*Aspergillus niger*	Glucose syrup production
	Rhizopus sp.	from liquefied starch
Glucose isomerase	*Actinoplanes missouriensis*	Isomerization of glucose into
	Streptomyces sp.	high fructose syrups
Lactase	*Saccharomyces* sp.	Hydrolysis of lactose in milk
	Kluyveromyces marxiamus	and whey
Lipase	*Aspergillus* sp.	Cheese and butter flavour
	Mucor sp.	modification. Fat and oil
	Rhizopus sp.	processing
Pectinase	*Aspergillus niger*	Extraction and clarification of
		fruit juices
Penicillin amidase	*Bacillus megaterium*	Synthesis of
	Escherichia coli	6-aminopenicillanic acid
		for manufacture of
		semisynthetic antibiotics
Protease (alkaline)	*Bacillus licheniformis*	Detergent and leather
		industries
Protease (neutral)	*Bacillus amyloliquefaciens*	Baking and brewing
Protease (acid)	*Endothea parasitica*	Cheese manufacture
	Mucor miehei	
Pullulanase	*Klebsiella pneumoniae*	Debranching starch in sugar
	'Bacillus	syrup manufacture
	acidopullulyticus'	

There are many ways to immobilize enzymes, the more common procedures involve (1) adsorption to an insoluble support of either organic or inorganic origin. Cellulose, dextran, nylon and bentonite are some of the many carriers that have been used. Attachment may be by physical adsorption, ionic binding or covalent bonding. (2) Entrapment methods in which the enzyme is localized within a polymer matrix are popular and include gel or fibre entrapment and microencapsulation in which the enzyme is enclosed within spherical semipermeable polymer membranes. (3) A simple but effective procedure is to immobilize the enzyme within the host cell by heat treatment or covalent cross-linking followed by pelleting the cells.

All these procedures are used by various companies to make immobilized glucose isomerase, by far the largest production of an immobilized enzyme. The immobilized pelleted or granulated enzyme is used in batch or continuous reactors for the production of high fructose syrups. Other immobilized enzymes include penicillin amidase, lactase, and glucoamylase for bulk manufacturing purposes but one of the indirect applications of this technology has been the development of enzyme-based analysers, enzyme probes and enzyme-linked immunosorbent assays.

Starch processing industry

Starch comprises two polymers; amylose, a linear molecule of 1,4-α-linked D-glucose residues and amylopectin, a branched molecule of short (20–25 residue) amylose chains joined at 1,6-α-branch points. Some 75–85% of starch is amylopectin depending on the source. Starch is located in water-insoluble granules in the tubers, seeds or stems of higher plants.

Traditionally starch was, and still is, hydrolysed to low molecular weight dextrins and glucose using acid, but enzymes have several advantages. Firstly, the specificity of enzymes allows the production of sugar syrups with well-defined physical and chemical properties. Secondly, the milder enzymic hydrolysis results in few side reactions and less 'browning'. Indeed, for the production of glucose syrups from starch enzymic hydrolysis is essential. After milling the source material, the granules are ruptured and the starch gelatinized by heat treatment. (Most enzymes are poorly, if at all, active on the native granule but the high energy costs of dispersing the starch by heat is prompting searches for amylases active on raw starch granules.) After cooling the gelatinized starch to about 80°C, the material is liquefied using an α-amylase (an endo-acting amylase that hydrolyses internal 1,4-α-linkages in amylose and amylopectin and rapidly reduces the viscosity of the molecules). The enzyme from *Bacillus amyloliquefaciens* has largely been replaced by that from *Bacillus licheniformis* for this purpose because of the higher temperature stability of the latter.

The liquefied starch is then 'saccharified' into low molecular weight malto-oligosaccharides with a combination of enzymes. Glucose syrups can be obtained by hydrolysis with glucoamylase (sometimes called amyloglucosidase; an exo-acting amylase that hydrolyses consecutive 1,4-α-bonds and the 1,6-α-bonds in dextrins to yield β-D-glucose). Sometimes a specific debranching enzyme is used to speed the reaction and improve the yield of glucose. Alternatively the liquefied starch can be hydrolysed with a fungal saccharifying α-amylase. These enzymes produce lower molecular weight oligosaccharides (glucose and maltose, from starch than the liquefying enzymes (maltopentase upwards). By careful balance of the ratio of glucoamylase to fungal α-amylase, high glucose syrup (30–50% glucose, 30–40% maltose) or high maltose syrups (30–50% maltose, 6–10% glucose) can be prepared. The former have high fermentability and are used in brewing and bread manufacture, the latter are less hygroscopic and have low tendency to crystallize and are used in jams and confectionery. Currently microbial versions of plant β-amylase that produce maltose from linear dextrins are entering the market and can be used in conjunction with debranching enzymes to make pure maltose syrups.

A major market for starch hydrolysates is for conversion into high fructose corn sweetners or syrups (HFCS). Glucose tastes about 70% as sweet as sucrose whereas fructose is about 50% sweeter and HFCS can therefore be used to replace sucrose syrups in foods and beverages. In the USA where corn starch is plentiful HFCS now dominates the carbohydrate sweeter market but within the European Economic Community high import tariffs have been introduced to protect the sugar beet industry. HFCS is produced by treating glucose syrups with xylose isomerase, an enzyme that catalyses the reversible reaction of xylose into xylulose or glucose into fructose. At equilibrium the product is about 55% fructose but most commercial preparations contain about 42% fructose. Some manufacturers use chromatographic separators to enrich up to 95% fructose. Immobilized glucose isomerase is by far the largest production of an immobilized enzyme with at least nine different manufacturers.

Proteases

Three classes of protease are produced commercially. Those with an alkaline pH optimum and serine residue at the active site are used in laundry detergents. Enzymes from *B. licheniformis* and alkaliphilic *Bacillus* strains are especially tolerant of the high pH of detergents and are reasonably temperature stable. Nevertheless, they are most effective when used in pre-wash soaking. Serine proteases have no metal ion requirement and so are resistant to the sequestering agents included in detergent preparations but they are sensitive to oxidizing agents such as hypochlorite. Enzyme based detergents improve laundering efficiency by hydrolysing coagulated proteins that may not be dispersed by the detergent. After a successful launch of enzyme washing detergents there was a setback in the 1970s owing to problems with allergic responses in workers and consumers. These have now been overcome by the use of dust-free enzyme preparations. Alkaline proteases are also used to dehair hides and to bate (soften) leather.

Metalloproteases have a pH optimum at or near neutrality and contain an essential metal atom, usually zinc. Manufactured from bacilli they are used to prevent protein hazes in beer and to reduce the gluten content of flour in the manufacture of biscuits and cookies.

Acid proteases have a low pH optimum and are more common in fungi than bacteria. Their principal application is to replace rennet (a proteolytic extract containing renin from the stomach of young calves) which is becoming more scarce. The coagulation of milk requires the partial hydrolysis of casein producing precipitation with little further hydrolysis. Most proteases will coagulate milk but their effect continues after coagulation causing over-ripening and a bitter taste. Acid proteases from *Mucor* and *Endothea* strains are acceptable rennet replacements and currently command about one-third of the total rennet market. In some traditional areas however there is reluctance to change to microbial rennets which has prompted several companies to clone the rennin gene into suitable hosts for the microbial production of 'calf rennet'.

Other enzymes

Several other enzymes that are produced in bulk are given in Table 8.1. This list is not exhaustive but includes the major products. Extracellular fungal cellulase preparations comprise mixtures of 1,4-β-endoglucanases and 1,4-β-exonucleases together with 1,4-β-glucosidases. They are currently used in fruit and vegetable processing but the potential for conversion of cellulose, the principal component of plant cell walls, into a fermentable hydrolysate is enormous and despite the recalcitrance of cellulose great advances are being made. Similarly hemicelluloses, which are substituted β-linked xylans are major components of plant cell walls. Xylanase preparations from fungi and streptomycetes have potential for the production of pentose sugars from this material.

Lactase (β-galactosidase) is secreted by many fungi and is used to hydrolyse lactose in ice cream to prevent crystallization and provide sweetness. A major application is the hydrolysis of lactose in milk whey which is produced at a world wide rate of about 1 million litres per day from cheese manufacture. Enzymic hydrolysis of the lactose into glucose and galactose using an immobilized lactase with simultaneous protein recovery is currently a promising method of upgrading this troublesome byproduct.

Penicillin amidase (acylase) catalyses the production of 6-aminopenicillanic acid from penicillin. This is a valuable precursor for the chemical synthesis of a variety of substituted penicillins with different clinical uses. The enzyme is derived from *E. coli* and immobilized before use.

Diagnostic enzymes

Enzymes in clinical laboratory practice have two distinct roles; they may be measured in body fluids as an indicator of disease or they can be used as chemical reagents for the estimation of

specific molecules in materials such as blood, plasma, serum and
urine, again to assist in disease diagnosis. In a few instances
enzymes are administered as therapeutic agents.

Enzymes as diagnostic reagents have the advantages of speed
and specificity over chemical procedures. In general they are convenient, reproducible and readily adaptable to automated systems.
Limitations include stability and the requirement for proper
storage and handling and interference by metal ions or proteins.
Nevertheless, even with new analytic techniques including high
performance liquid chromatography and monoclonal antibodies,
diagnostic enzymes continue to enjoy an increasing market.

The most common clinical laboratory measurement is glucose
in both urine and blood. A linked assay employing glucose oxidase
and peroxidase is used for this (Fig. 8.1) although there are variations. Kits are available from most manufacturers and various
automated analysers have been developed. Glucose represents
one of the earliest rapid analyses in the form of 'Clinistix' a dip
and read reagent system in which the enzymes are impregnated
in filter paper mounted on a plastic strip and immersed in the
urine. Generation of a blue colour indicates glucose.

(a) $\text{Glucose} + O_2 + H_2O \xrightarrow[\text{oxidase}]{\text{Glucose}} \text{gluconate} + H_2O_2$

$\quad H_2O_2 + \text{chromogen} \xrightarrow{\text{peroxidase}} \text{colour} + H_2O$

(b) $\text{glucose} + \text{ATP} \xrightarrow{\text{hexokinase}} \text{glucose-6-phosphate} + \text{ADP}$

$\quad \text{glucose-6-phosphate} + \text{NADP}^+ \xrightarrow[\text{dehydrogenase}]{\text{G-6-P}} \text{gluconate-6-phosphate} + \text{NADPH} + H^+$

Fig. 8.1. Two common procedures for measurement of glucose in blood and
urine (a) a linked colourimetric assay in which a variety of chromogens can be
used (b) a linked assay in which the reduction of NADP^+ is followed
spectrophotometrically at 340 nm.

Prepacked kits are now available for at least 20 clinically
important compounds with cholesterol, glucose, urea, uric acid
and triglycerides as the most common. Assays involve either single
enzymes or linked multienzyme systems, the most common
end points being the conversion of NAD^+ to NADH and the formation of hydrogen peroxide measured with peroxidase and a
chromogen.

Enzyme analysers and electrodes

Enzyme analysers are based on the flow-through auto analysers common in clinical chemistry and use diagnostic enzymes immobilized within nylon tubing. For example, for glucose estimation hexokinase and glucose-6-phosphate dehydrogenase have been immobilized in a nylon coil by Technicon Corporation and glucose concentration is determined by the rate of formation of NADH using an in-line spectrophotometer. Similarly, creatinine iminohydrolase, an enzyme that releases ammonia from creatinine has been immobilized in a nylon coil and the ammonium ions released drive a glutamate dehydrogenase catalysed conversion of NADH to NAD^+ which can be monitored at 340 nm. Other nylon coil based systems include determination of urea, uric acid and cholesterol.

Enzyme electrodes comprise the immobilized diagnostic enzyme in conjunction with an electrochemical sensor. The substrate diffuses through a membrane, reacts with the enzyme and the products are detected by the electrode. A variety of electrodes have been used including pH electrodes, ion-selective electrodes, oxygen electrodes, gas-sensing electrodes, thermistors and light-sensing devices. Although over 50 substrate electrodes have been described in the literature only that for glucose based on immobilized glucose oxidase and a platinum electrode to detect the resultant hydrogen peroxide is marketed on a large scale. One reason for this is problems with poor selectivity of the transducer. For example urea and creatinine electrodes use immobilized urease and creatinine iminohydrolase respectively to generate ammonium ions. However, ammonium electrodes are not very specific and respond to other monovalent cations thus causing interference. Nevertheless, the extreme utility of, and large markets for, enzyme-based probes will doubtless lead to the development of improved systems.

Enzyme immunoassay

Enzyme immunoassay techniques are rapidly replacing radioimmunoassay in clinical chemistry, since they offer high levels of sensitivity and reproducibility together with broad application and the possibility of automation. Moreover, there is no contact with radioactive material. Enzyme immunoassays can be broadly classified as heterogeneous and homogeneous.

Heterogeneous assays employ a solid support, usually a plastic tray or tube to separate free from bound enzyme-labelled molecules in a procedure termed enzyme-linked immunosorbent assay (ELISA). The simplest procedure, the competitive enzyme–immunoassay is shown in Fig. 8.2(a) and is analogous to the classical radioimmunoassay. The antigen is labelled with an enzyme such as alkaline phosphatase or β-galactosidase and competes with antigen in the sample for sites on the immobilized antibody.

After removing excess substrate by washing, a chromogenic substrate is added and the enzyme reaction monitored. The amount
of enzyme-labelled antigen bound is inversely proportional to the
concentration of antigen in the sample. The direct sandwich
method involving enzyme labelled antibody is shown in Fig. 8.2(b)
and other variations exist.

Homogeneous enzyme-immunoassay involves attaching the
marker enzyme to the antigen and monitoring a depression in
enzyme activity when the antigen combines with antibody
(enzyme multiplied immunoassay technique, EMIT). Thus the
enzyme activity depends on the concentration of free antigen in
the sample which competes with the antigen–antibody complex
for a limited concentration of antibody (Fig. 8.2[c]). Other
homogeneous techniques have been developed as have the
highly sensitive immunometric assays employing monoclonal
antibodies.

Monoclonal antibodies have the advantage over their polyclonal counterparts in that a specific immunoglobulin line is produced that reacts with a single epitope within the antigen. This
high degree of specificity has enabled the application of immunometric techniques (Fig. 8.2[d]) in which a solid phase monoclonal
antibody is reacted with the antigen such that the immune complex
is formed with a specific site. This is then reacted with an excess
of labelled monoclonal antibody to a second epitope on the antigen
and, after washing, the label (radioactive, enzyme, etc.) may be
quantified. Higher sensitivity and shorter reaction times make this
a particularly valuable technique for the quantitation of very low
concentrations of hormone in body fluids.

Therapeutic enzymes

Enzymes have found value as therapeutic agents in a variety of
applications. One of the most successful examples involves the
treatment of certain neoplasmas including acute lymphocytic
leukaemia with asparaginase. These tumours lack the capacity to
synthesize asparagine and, by maintaining low levels of the amino
acid in the body, the neoplastic cells are unable to grow while
normal cells are unaffected.

An obvious use of enzymes is in the treatment of enzyme
deficiencies but a serious problem has been the effective delivery
of the enzyme to the appropriate tissue. This is now being tackled
using liposomes with some success.

A particularly important application of therapeutic enzymes
is parenteral administration of fibrinolytic enzymes. Cardiovascular diseases such as heart attack and strokes usually arise
from the obstruction of a blood vessel by a clot. The natural
formation and dissolution of fibrin which constitutes the clot is

Immunological reaction

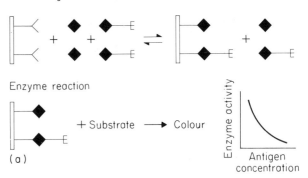

(a)

Fig. 8.2(a). Some common forms of enzyme linked immunoassay. The competitive assay in which an immobilized IgG is reacted with sample containing the antigen (◆) and a preparation of antigen linked to enzyme (◆-E). After washing, the amount of enzyme linked antigen bound to the immobilized IgG can be estimated colorimetrically and is inversely proportional to the concentration of antigen in the sample.

Immunological reaction

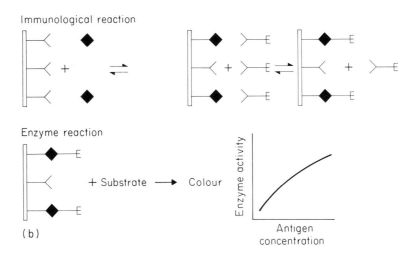

(b)

Fig. 8.2(b). The direct sandwich assay involves reacting the sample containing antigen (◆) with immobilized IgG. This complex is then reacted with an IgG enzyme formulation (◆-E) and the amount bound can be determined colorimetrically. In this procedure the antigen concentration is directly proportional to the enzyme activity.

carefully balanced in the blood stream and tissues. The process of fibrinolysis involves the activation of plasminogen into the proteolytic enzyme plasmin which acts on fibrin. Treatment therefore involves the preparation and administration of an activator of plasminogen. Typical plasminogen activators include urokinase, a protease from urine, and streptokinase which is synthesized by β-haemolytic streptococci. However, the systemic

High antigen concentration   Low antigen concentration

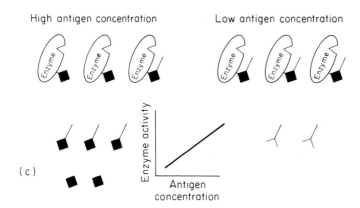

(c)

Fig. 8.2(c). Homogeneous enzyme-linked assays do not involve immobilized IgG. Instead the antigen is bound to an enzyme. This is reacted with free antigen in the sample and IgG. Binding of the IgG to the enzyme complex inhibits enzyme activity. Thus if there is a high concentration of antigen in the sample it will bind the IgG and the enzyme activity will be high. If there is a low antigen concentration in the sample, the IgG will bind to the enzyme complex and the enzyme activity will be low.

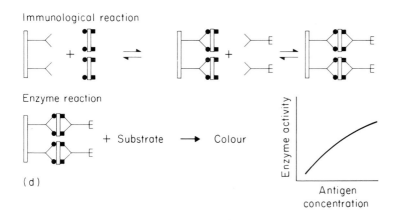

Immunological reaction

Enzyme reaction

+ Substrate ⟶ Colour

(d)

Fig. 8.2(d). The immunometric assay is similar to the direct sandwich assay but exploits the epitope specificity of monoclonal antibodies. The antigen is reacted with an immobilized IgG through one epitope and a second IgG linked to an enzyme is complexed with the bound antigen. Enzyme activity is directly proportional to antigen concentration even at exceptionally low levels.

use of these agents requires large doses and freedom from antigenicity is important. Thus, streptokinase, which is strongly antigenic, is of little use and urokinase, of human origin and therefore nonantigenic is only found in trace amounts in urine. Nevertheless urokinase is used in the treatment of a number of thrombic disorders.

With the capability to culture human cells *in vitro* it became possible to produce the tissue-type plasminogen activator (t-PA) that, unlike urokinase preferentially activates the plasminogen in clots. Some cancer cells produce large amounts of t-PA but, because of the possibility of release of cell-transforming factors the material is not used clinically. However, more recently large amounts of t-PA have been derived from human embryonic lung cells and this has the potential for development into commercial production.

Chapter 9. Foods and beverages

This is certainly the oldest biotechnology and in economic terms remains the most significant application of industrial microbiology.

Beer

The origin of beer brewing is lost in antiquity but archaeological evidence shows that brewing was practised in Babylon in 6000 BC. Beer brewing was a domestic activity until in medieval Europe large scale production was concentrated first in monasteries and then in commercial breweries.

Beer is produced by fermentation of an extract of malted cereals, preferably barley. Aromatic herbs were often added for additional flavouring and hops have become the standard flavouring of modern beers. Cereal grains contain little fermentable sugar; the carbohydrate is largely in the form of starch which few yeasts are able to utilize. Therefore as a preliminary to fermentation, the grain is moistened to encourage germination, during which starch- and protein-hydrolysing enzymes are formed, to provide sugars and amino acids as nutrients for the embryo plants. At an appropriate stage of development the grain is heated just sufficiently to kill the embryo plants; malt kilns generally operate at around 65–80°C, according to type of malt required. The heating process both dries the malt, to permit storage, and improves the flavour, but does not inactivate the enzyme content of the grain. Although modern technology has improved the process of malting, it is essentially the process developed 8000 years ago, and it is interesting that the complex production of fermentable material from grain was discovered so early in human history.

In modern brewing the malt is ground and extracted with hot water ('mashed'); often ground unmalted cereal is added at this stage. Since sufficient enzyme activity remains in the malt after kilning, the starch of the 'adjunct' grain is hydrolysed to fermentable sugar during mashing. Mashing temperatures vary, but are generally in the range 50–80°C, and the hydrolytic enzymes of malt remain sufficiently active to complete the hydrolysis of starch.

The sugary extract, wort, drained from the mash tun is clarified by the husk particles functioning as a filter. The mash is sparged with either one or two further batches of hot water to ensure maximum extraction of nutrients, and the collected 'sweet wort' is boiled with hops, primarily to extract flavour. Boiling also sterilizes the wort and inactivates enzymic activity. Again, draining off the wort is effectively a process of clarification; the precipitate of protein, tannins and phosphate produced by boiling is filtered off by the bed of hop debris and a clear 'hopped wort' is obtained. After cooling, the wort is inoculated with a suitable strain of yeast. Normally air is injected immediately before inoculation to improve yeast growth.

The main products of fermentation are ethanol and CO_2, but small amounts of numerous by-products of yeast growth are also formed, as important contributors to the flavour and aroma of the beer (Table 9.1). Organic acids, alcohols and esters are especially important in this respect. During fermentation the pH falls from the initial level of 5.0–5.2 to pH 3.8–4.0.

Table 9.1. Products of yeast fermentations.

Alcohols	Acids	Esters	Others
Ethanol	Acetic	Ethyl acetate,	CO_2
n Propanol	Lactic	and other esters	Acetaldehyde
Butanols	Pyruvic	of acid and alcohol	Diacetyl
Amyl alcohols	Succinic	fermentation products	H_2S
Phenylethanol	Caproic		
Glycerol	Caprylic		

Ethanol, glycerol and CO_2 are the principal products of the fermentation pathway.
Other compounds, even in low concentration, contribute to flavour and aroma.

The yeast population grows approximately 8-fold during the fermentation, limited partly by the falling pH and rising ethanol concentration, but mainly by its inability to grow indefinitely under anaerobic conditions. Although it is customary to aerate the wort during cooling after hop-boiling, that oxygen is rapidly consumed and aeration later in the fermentation is unacceptable because of its effect on flavour.

At the end of the fermentation, with no further evolution of CO_2, the yeast should settle out in the nonturbulent conditions. Clarification is accelerated by chilling, but a good yeast strain will spontaneously flocculate into clumps at the end of fermentation and settle out rapidly. Further clarification by fining agents (especially the collagen protein isinglass), filtration or centrifugation is practised if necessary.

The general description given above, and illustrated in Fig. 9.1, is necessarily vague, to allow for the great variation in practice between breweries. In particular, the methods of preparation of the two main types of beer, ale and lager, are traditionally different

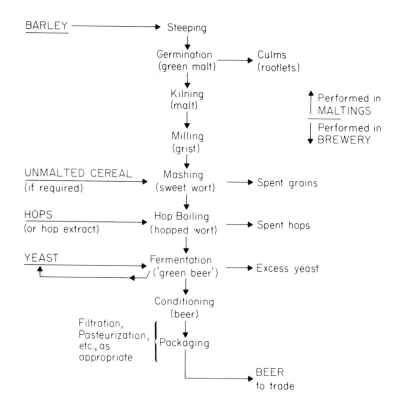

Fig. 9.1. Flow diagram of beer production.

(including different malts, hops and yeasts), but in recent years the differences have become less obvious.

Beer of the ale type is traditionally fermented by 'top' strains of *S. cerevisiae,* so termed because a proportion of the yeast rises to form a thick 'yeast head' on the surface of the fermenter. The yeast head is collected during fermentation to provide the inoculum for a following fermentation. Beer of the lager type originated in Bavaria in medieval times and has subsequently become the dominant type world-wide. Malt for lager was kilned at a lower temperature (as we now recognize, to provide greater enzyme activity) and has a lighter colour. The yeast, formerly identified as a separate species (*S. carlsbergensis*) did not form a yeast head and was harvested from the bottom of the fermentation vessels, where it settled at the end of the fermentation. 'Bottom yeast' grows well at lower temperatures than 'top yeast'. The main, or primary, fermentation was customarily at 8–10°C, followed by a prolonged secondary fermentation, of up to 3 months, at 0°C which improved the flavour and stability of the beer.

Modern techniques have blurred the differences between the two types of beer, in particular in that the mashing systems are less distinctive than formerly, and 'bottom yeast' is often used for both ale and lager. This last change was an incidental development from the introduction of enclosed fermentation vessels, in which the space required for yeast head represents a loss of useful fermentation volume. Modern cylindroconical fermenters (Fig. 9.2) generate a vigorous natural movement of the yeast through the fermenting wort: rising with CO_2 bubbles and sinking by cooling at the walls of the vessels. The fermentation is faster, and protected from microbial contamination and the vessels are easily equipped with automatic cleaning and sterilizing equipment. Nevertheless, many breweries continue to operate open rectangular vessels of the traditional form, especially for the production of cask conditioned beers.

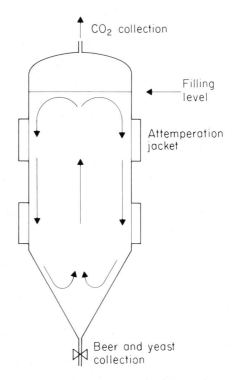

Fig. 9.2. Cylindroconical fermentation vessel for beer fermentations, showing pattern of mixing generated within the vessel.

Only a small amount of beer is now conditioned in cask, where the CO_2 content of the final beer is generated by a secondary fermentation. Isinglass finings, added at filling, ensures a clear beer by coagulating the yeast once the fermentation-in-cask is over. Otherwise, modern practice is to perform these processes in bulk, in the brewery, under consistent conditions. All beers require a period of conditioning after fermentation; in the case of ales only a few days chilling is required to improve the flavour of the 'green beer' drawn from the fermenters. At the same time

CO$_2$ is injected as required, and the beer is clarified by filtration and pasteurized for longer shelf life.

93
Chapter 9
Foods and
beverages

Wine

The word wine without qualification specifically means fermented grape juice, but other fruits are also used for wine production. Wine fermentations may be 'natural' or 'artificial' according to whether the natural yeast flora or an artificially grown culture yeast is used.

Natural wine fermentation is now largely confined to European wine-producing areas, particularly France. Grapes develop a microbial flora during development, and on pressing, these organisms inoculate the juice. Micro-organisms of the pressing equipment and other sources in the winery also contribute an inoculum to the juice. The juice, or must, is first treated with sufficient sulphite to eliminate undesirable yeasts, moulds and bacteria, but not so much as to harm the fermentation yeasts, which fortunately are more resistant to SO$_2$ or sulphite. During the fermentation a succession of yeasts develops; it is rare for only one strain of yeast to occur. Species of *Saccharomyces,* including *S. cerevisiae,* are involved, but also various species of other yeast genera, for example *Kloeckera, Kluyveromyces, Torulaspora* and *Zygosaccharomyces.* Each yeast contributes its own spectrum of flavour compounds, and as the yeasts involved vary from year to year, so too do the flavour and aroma ('bouquet') of the wine. The characteristics of the wine are due partly to the grapes, and partly to the yeast. Although the grape variety (or varieties) used is constant, the climate each year affects sugar content, acidity, etc. and so influences the flavour of each year's production. It is possible on rare occasions that the yeast flora introduced naturally is unsuitable, in which case a specially grown culture of wine yeast is added but the wines which result are generally judged to be of poorer quality.

However, it is standard practice in most wine-producing areas to use a pure yeast culture or a mixture of pure cultures for all fermentations. All microbial flora of the grape juice are eliminated by addition of greater amounts of SO$_2$ or sulphite. Pasteurization of grape juice is used in some areas for production of cheaper wines, but is generally avoided because of its effect on flavour. Wine produced by 'artificial' fermentation is of consistent flavour from year to year, apart from the minor variations caused by climatic effects on the grapes themselves. Table 9.2 indicates the most important properties of wine yeasts for 'artificial' fermentations, or for rescuing 'natural' fermentations in difficulty.

The tannins and pigments of grape skins are important. Red wines are produced from black grapes, the skins of which are left

Table 9.2. Important properties of wine yeasts.

1	High ethanol production (up to 15% v/v for some wines)
2	Resistance to high sugar concentrations (up to 30% w/v)
3	Resistance to sulphite
4	Resistance to tannin (especially for red wines)
5	Resistance to ethanol (capability for growth in moderate ethanol concentrations: this is important for yeasts used to 'rescue' faulty 'natural' fermentations
6	Wide temperature range for growth (e.g. 4–32°C)
7	Low production of volatile acidity (measured as acetic acid)
8	Good flavour production (compounds in Table 9.1)
9	Fermentation under pressure ⎫ for sparkling wines
10	Firm deposit ⎭

in contact with the fermenting juice. Grape skin pigments are ethanol-soluble, and extracted as fermentation progresses. Also tannins are extracted, conferring an unpleasant bitterness in young red wine that mellows by chemical reaction and precipitation as the wine ages. White wines are produced from either green or black grapes, but the skins must be removed before commencement of fermentation. Therefore white wines have lower tannin or polyphenol content than red wines, which may affect their stability in certain circumstances (for example low ethanol or sugar content) unless protected by SO_2 or other preservative.

The progress of the wine fermentation is essentially the same as the beer fermentation, but normally over a longer time and producing higher concentrations of ethanol, typically 10–12%. Postfermentation treatments vary widely, according to the locality and type of wine, but normally prolonged storage in cask is necessary for satisfactory clarification and maturation before bottling.

Secondary fermentation

Production of Champagne and sherry wines involves a secondary fermentation during the maturation process. In the 'Méthode Champenoise', CO_2 is dissolved under pressure by secondary fermentation in the bottles. Therefore strong bottles are required, capable of withstanding the 6 bar pressure that can develop. Bottles are filled with new wine containing 1% fermentable sugar. Unfermented grape juice is added if necessary to provide the necessary sugar. The yeast remaining in the wine continues the fermentation in the sealed bottles, and over a period of many months the bottles are rearranged at intervals until vertical, stopper downwards, and all of the yeast cells produced during the secondary fermentation have collected on the cork. The yeast is removed by freezing the neck of the bottle to provide a temporary plug of ice; the cork with attached yeast is removed and a new cork is fitted and secured by wire to seal the now clarified wine. (**Warning:** the high pressure makes this a dangerous process, NOT to be attempted by amateurs.)

This is a labour-intensive and therefore expensive process, and in some areas a cheaper sparkling wine is produced by carrying out the secondary fermentation in stainless steel tanks. After the yeast has settled out, the clear wine is bottled against a counter-pressure of CO_2. Note that in both types of sparkling wine the CO_2 is dissolved over a long period of secondary fermentation. A cheap, but inferior, sparkling wine can also be produced by injection of CO_2, as in the manufacture of aerated soft drinks. Such a product is recognizable by the rapid loss of CO_2 from the wine in the glass.

A different type of secondary fermentation is involved in the production of sherry wines. In the Jerez region of Spain, which by mispronunciation has provided the name Sherry, the wine is matured in warehouses (*bodegas*) in a solera system, in effect a stack of wine casks. After drawing off half of the content of the lowest (oldest) cask for bottling, that cask is topped-up from the level immediately above. Repetition of this process at higher levels leaves space in the highest (youngest) cask for new wine. The light colour and dry flavour of Fino Sherry is due to the film (*flor*) of surface yeast growth. Various species of film-forming yeasts may be involved, most commonly *Saccharomyces cerevisiae, Torulaspora fermentati* and *Zygosaccharomyces rouxii.* These yeasts are capable of growing in the high ethanol concentrations (over 12%) of the new wine and increasing the level to approx. 18%, producing not only ethanol but also acetaldehyde and glycerol as particularly important flavour compounds. The yeast film prevents access of air to the wine, which therefore remains unoxidized, and light in colour.

The darker, sweeter, sherries are also matured in a solera system, but the primary fermentation is stopped at an appropriate level of residual sugar by addition of sufficient brandy to inhibit further yeast growth. During maturation the reaction with air, both at the wine surface and through the wood of the cask, develops a darker colour and oxidized flavour.

Yet another type of secondary fermentation is of value in the wine industry to reduce the acidity of wines. However, the malolactic fermentation is not strictly a secondary fermentation, since it may occur either during the primary fermentation or during maturation. The conversion of malic (with two carboxyl groups per molecule) to lactic acid (with only one carboxyl group) effectively halves the acidity, and is often a desirable by-product of 'contamination' by *Leuconostoc oenos*, the usual bacterium involved (Fig. 9.3). Deliberate inoculation of *L. oenos* is shunned by the purists, but is done in some wine-producing areas,

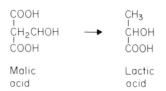

Fig. 9.3. Malolactic fermentation (*Leuconostoc oenos*).

either late in the primary fermentation or in the early stages of maturation.

Many fermentable fruit juices other than grape are also used in the production of wine, and normally the process is similar to grape wine production in all aspects. Cider and perry, from apples and pears respectively, also are produced by a similar process, and both naturally and artificially inoculated products exist, as with wine production. In a few parts of the world the causative organism of natural alcohol fermentation is the bacterium *Zymomonas,* rather than a yeast. Although the metabolic route is different (the Entner–Doudoroff pathway in *Zymomonas*) the main end products are ethanol and CO_2, as with yeast. However, the different range of minor byproducts of fermentation causes *Zymomonas* to be regarded as a troublesome source of off-flavour in the beer, cider and wine industries.

Distilled spirits

In most areas of the world, local types of distilled spirit have been developed. Of these, brandy, gin, vodka and whisky are the main acceptable types, available world-wide. Brandy without qualification specifically means distilled grape wine, but brandies are also produced from other fruits, especially apple, cherry and plum. Rum is the distilled spirit of fermented sugar-cane juice or molasses, and whisky (in Scotland) or whiskey (elsewhere) is the distilled spirit of fermented cereals. Vodka is diluted pure spirit from any source, usually cereal, but in some countries potato or other starchy root crops are permitted. The same spirit can be prepared as gin by a final distillation with flavourings, usually including juniper, caraway and orange, or for a cheaper version, by addition of these flavourings after distillation. Addition of flavourings, before or after distillation, is also a method of production of liqueurs.

The methods of production of different distilled spirits have sufficient in common that whisky production can be used as an example. In Scotland two types of whisky are produced: malt whisky, in which only barley malt is permitted, and grain whisky, in which the greater part of the fermentable material is unmalted cereal, normally maize or wheat, and malted barley is used mainly as a source of the necessary hydrolytic enzymes. Very little of the grain whisky produced is drunk in that form: virtually all is blended with malt whisky to produce the various brands of Blended Scotch Whisky.

Production of grain whisky (Fig. 9.4) requires a malt of high enzymic activity. 'Green malt' (i.e. unkilned) is most effective, although with the penalty of short storage time because of the high moisture content. Wheat is now increasingly used as the most economical form of starch, and the grain is crushed and steamed to gelatinize the starch before mixing with the crushed malt. The mash, or 'wash', is discharged directly to the fermentation vessels without clarification or boiling (although the larger particles of grain debris are normally filtered off) so that the full activity of the malt enzymes is available during the fermentation, yielding additional fermentable sugar. Distillers yeast is not itself diastatic, but is capable of utilizing the sugars glucose, maltose and maltotriose, like brewer's yeast, and also maltotetraose.

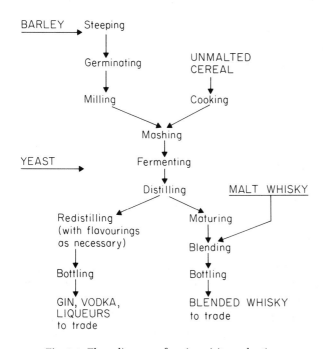

Fig. 9.4. Flow diagram of grain spirit production.

Distillery fermentations are not normally attempered, and the vigorous fermentation causes the temperature to rise to as much as 35°C. The fermentation is complete within 25–40 h and the product, including grain debris and yeast, is distilled in continuously operating Coffey stills (*see* Fig. 7.2). Ethanol and higher alcohol fractions are collected separately, the higher alcohols being sold as industrial solvent. The ethanol is diluted to approx.

65% for maturation in wooden casks for at least 3 years (the minimum legal period) but normally not less than 5 years, and finally blended with malt whisky for sale. Wooden casks are essential for oxygen transfer, but only a restricted range of oak varieties is used, to avoid complications of extraction of undesirable flavour compounds from the wood.

The preparation of the mash for malt whisky production is essentially as in a brewery, but there is no stage equivalent to hop-boiling and the enzymic activity of the malt is retained throughout the fermentation. A high level of diastatic activity is not required, so green malt is unnecessary. Traditionally the malt is dried in peat-fired kilns, adding a smoky and phenolic aroma which persists to the final whisky.

As in the grain whisky distilleries, there is no attemperation of the fermentation. After fermentation, distillation is performed batchwise in pot stills (*see* Fig. 7.1). The efficiency of distillation is lower, both in terms of ethanol concentration and separation of aromatic by-products of fermentation, therefore a stronger-flavoured distillate is produced. Stills vary widely in shape, but since the shapes of both still and condenser affect the quality of the spirit, replacement stills are always exact copies of their predecessors. Copper construction is essential for acceptable flavour; no other metal is sufficiently reactive with unwanted sulphur-containing byproducts of fermentation.

Subsequently the spirit, adjusted to 60–70% ethanol, is matured in oak casks for a period unlikely to be less than 8 years, and sold either as a Single Malt Whisky or for blending.

A proportion of grain spirit production is subjected to further distillation to produce spirit with the least possible content of higher alcohols, for gin and vodka production. Filtration through activated carbon may also be used to remove compounds other than ethanol from vodka spirit. Best-quality gin undergoes a final distillation with 'botanicals' to extract volatile aroma compounds from these flavourings; cheaper products are flavoured with essences. Maturation in casks adversely affects the flavour of gin and vodka, which are therefore bottled immediately after the final distillation and dilution with pure water (to avoid precipitation of salts) to sale strength.

Baker's yeast

Beer fermentations produce an excess of yeast which throughout most of civilized human history was used as an ingredient in bread. Yeast from wine and spirit fermentation was unsuitable, having been metabolically damaged by the high final ethanol concentration. By the early 19th century the breweries could not supply sufficient yeast and brewery-type malt fermentations, but omitting the unnecessary hops, had to be operated solely to supply sufficient yeast for the greatly expanded baking industry. This inefficient system was improved throughout the 19th century, first by using cheaper molasses instead of malt, and, following Pasteur's discoveries on the nature of aerobic and anaerobic growth, by aeration of the molasses medium. Indeed, it was by such improvements that modern fermentation processes now exist: the development of an efficient process for manufacture of baker's yeast pioneered the aerobic fermentation processes that would become so important in the mid-20th century.

Aerobic growth of yeast introduced three main problems: (1) yeast and bacterial contaminants, previously unimportant, were stimulated by the aerobic conditions; (2) vigorous aerobic growth generated excessive heat; and (3) in the turbulent conditions of active growth the yeast no longer formed a head which could be skimmed off, or readily settled out by sedimentation of flocculation.

This stimulated the development of closed fermentation vessels, capable of sterilization, or at least elimination of troublesome contaminants, by heating to 100°C, and equipped with water jackets for temperature control and a supply of sterile air for aeration. Although initially treated unsatisfactorily by bubbling through chemical sterilants (for example NaOH), air was most effectively sterilized by filtration. Filtration methods, previously not applied knowingly to large scale processes, were also developed for separation of the yeast, now achieved by rotary vacuum filter (see Fig. 7.4). Subsequent developments included the use of pure yeast cultures specifically bred for the necessary properties for the baking industry (Table 9.3), which were grown in smaller fermenters as inoculum for the production fermenters. This became standard practice in all modern fermentation processes. The

Table 9.3. Comparison of properties of baker's yeast and yeast for industrial alcohol production.

Baker's yeast	Alcohol yeast
1 Good storage stability	1 Rapid fermentation
2 Good survival of drying and reconstitution procedures (applies only to dried yeast)	2 Tolerance of high initial sugar concentration
3 Rapid fermentation of the low sugar concentration in bread dough	3 Tolerance of high final ethanol concentration
4 Constitutive fermentation of maltose (since grown in sucrose medium)	4 Low production of byproducts of metabolism (higher alcohols, esters, etc.)
5 Osmotic tolerance (since only a small amount of water is used in dough mix, salt concentration is high)	5 Growth at as high a temperature as possible

efficiency of yeast growth was improved by maintaining a consistently low concentration of sugar in the medium. Above 1% fermentable sugar, even with vigorous aeration, yeast ferments sugars to CO_2 and ethanol, and is unable to complete oxidative metabolism to CO_2 and water. Therefore, with knowledge of this Crabtree effect, an incremental feeding regime of molasses was applied, only slowly supplying molasses at the start of the fermentation to a sterilized basal medium containing N and salts. As the yeast population increased, concentrated molasses was introduced more and more rapidly to maintain an adequate, but low, nutrient concentration. The end of the fermentation is dictated by the oxygen supply to the yeasts: when the yeast concentration reaches the maximum that can be efficiently aerated, the fermentation is stopped and the yeast culture harvested. Normally this stage is reached in 18–20 h.

Baker's yeast is sold in two forms: moist and dry. Yeast cake from the rotary filter may be packaged in that form: although requiring refrigeration during storage and of maximum effective shelf life 2–3 weeks, it has the advantage of easy mixing with bread dough. Yeast dried under partial vacuum at 55°C remains active for at least 1 year at room temperature, provided it is kept dry, but has to be carefully reconstituted to avoid osmotic damage. However, recent improvements in yeast strains and in drying technology have resulted in a more easily reconstituted yeast.

Milk products

In considering fermented milk products we revert to traditional fermentations. Raw milk is rapidly spoilt by microbial activity and throughout the world fermented milk products have been developed with longer shelf life. Such products invariably involve homofermentative lactic bacteria, which ferment lactose largely to lactic acid but in addition produce traces of other, minor, byproducts of metabolism which contribute to flavour. By its low pH, soured milk is protected from proteolytic bacterial spoilage and it is easy to imagine how the range of fermented milk products would have been discovered by dairy-farming communities. Yoghurt and cheese are two particularly important types of fermented milk product.

Yoghurt

Yoghurt is prepared from artificially concentrated milk, traditionally by boiling but more economically in modern practice by addition of skimmed milk powder to pasteurized skimmed milk to bring the nonfat solids content to approximately 10%. Traditional inoculation by a sample of a recently prepared yoghurt of satisfactory quality has, for largescale production, been replaced by inoculation with a culture of the appropriate organism or mixture of organisms. *Lactobacillus bulgaricus* and *Streptococcus thermophilus* are two suitable homofermentative species often used together, with the useful property of producing thixotropic polysaccharides which improve the consistency of the yoghurt. Fermentation at 42–45°C for 4 h generates sufficient acidity (pH approx. 4.0) for satisfactory shelf life of natural yoghurt. Addition of fruit, even of good microbiological quality, represents a possible hazard, and particularly by introduction of yeasts which could generate unwanted ethanol and CO_2. Fruit yoghurts are customarily protected by addition of sugar, to provide a lower pH and higher osmotic pressure, although incidentally also sweetening the product. Even so, fruit yoghurts, pasteurized or not, must be stored at refrigerator temperatures to achieve adequate shelf life.

Fermentation and associated equipment in modern yoghurt production is stainless steel, and of capacity up to 10,000 litres in the largest plants. Heterofermentative lactic bacteria, oxidative and fermentative yeasts, and moulds, are important contaminants which spoil the quality of the product, but fortunately without hazard to health. The most serious industrial contaminants are bacteriophages specific for the culture bacteria. Therefore all sterilization and transfer operations must be designed to avoid introduction of phage; fortunately the cultures grow without aeration, which would otherwise be a potential source of 'phage contamination'.

Cheese

Cheese represents a method for preservation of milk protein without dangerous proteolytic microbial spoilage. Traditionally the casein and other minor proteins are coagulated by the enzyme rennin extracted from calf stomach but shortage of supply requires that other weakly proteolytic clotting enzymes of animal or microbial origin must be used if necessary. Rennin (rennet) has a low pH optimum, and substitutes are normally of similar properties, therefore the first stages of cheese manufacture are pasteurization of the milk and inoculation with a suitable starter culture to generate the necessary acidity. Various mesophilic or thermotolerant lactic bacteria are used, according to the tradition for each type of cheese. As for yoghurt production, cultures are prepared under conditions which avoid spoilage by heterolactic bacteria, yeasts, moulds or bacteriophage. Although various *Streptococcus* and *Lactobacillus* species are involved in the manufacture of most cheeses, other genera may be involved, for example *Propionibacterium* in the production of Swiss cheeses with gas bubbles.

The enormous variety of cheeses arises partly from the range of starter culture organisms available, originally present in the unpasteurized milk used for cheese manufacture but now propagated as commercial cultures. The differing proteolytic and lipolytic activities, and the amounts and range of minor metabolic products, all influence flavour and texture of cheeses from different bacterial strains. Secondly, the treatment of the rennin-induced curd varies according to the type of cheese: the extent of pressing to expel water, heating or cooling to control microbial or enzymic activity, addition of salt to modify the microbial flora by osmotic effects, and mechanical manipulation to modify texture, all influence the final product. Thirdly, certain varieties of cheese are deliberately infected by mould cultures: either surface growth of *Penicillium camemberti,* as in Camembert, Brie and similar cheeses, or also in the depth of the cheese along the inoculation-needle track, as in Stilton, Danish Blue, Gorgonzola and Roquefort cheeses, seeded with *P. roqueforti.* Both *Penicillium* species modify cheese flavour by their proteolytic and lipolytic activities.

Although bovine milk is used for most cheeses, milk from other domesticated animals is used for certain varieties (for example Roquefort from ewe's milk), with obvious effect on flavour. The additional preservative effect by smoking has been found to be useful and such smoked cheeses continue to be produced on a small scale.

Cheese manufacture itself is rightly considered one of the older biotechnologies, comparable with brewing and wine-making. Modern biotechnological interest in the cheese industry has produced better starter cultures, in terms of flavour, acid production and phage resistance, and rennin substitutes of microbial origin. Additionally, the whey left after collection of the curd is a nutrient-rich byproduct which may be used as a substrate for production of food yeast or ethanol by *Kluyveromyces marxianus,* which is capable of utilizing the lactose, the principal sugar. Whey also provides a substrate for lactic acid production, by *Lactobacillus* spp.

Organic acids

Until 1923, citric acid was prepared from the juice of citrus fruits, particularly lemons and limes, specifically grown for the purpose. In that year a fermentation system was developed using a strain of *Aspergillus niger,* now known to have a defective citric acid cycle, which quantitatively oxidized glucose or sucrose to citric acid (Fig. 9.5).

In practice the yield was not 100%, but by re-use of already grown mould mycelium, thereby avoiding loss through synthesis

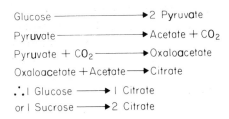

Fig. 9.5. Biosynthesis of citrate (*Aspergillus niger*).

of new cell material, high ($> 90\%$) yields were obtained. The original production system used large shallow trays of porcelain (since no acid resistant metal was then available), up to 1 m square, stacked in dust free rooms (Fig. 9.6). Diluted molasses, with supplementary nutrients as required, is the most convenient medium, and is sterilized by brief heating to 100°C. This is sufficient, since immediately on cooling the medium is acidified to pH 2.5. The low pH prevents contamination, since few fungi other than *A. niger* can grow, but also prevents the production of oxalic acid, which would occur at pH 5 or higher. After 6–7 days at 30°C the sugar has been converted to citric acid, which is recovered by drawing off the spent medium and treating with lime or chalk. The precipitated calcium citrate, treated with sulphuric acid, yields a crude citric acid which is adequate for many of the industrial applications of the compound. Recrystallization produces a better quality food grade acid for use as a stabilizer for frozen foods, or as an ingredient of soft drinks.

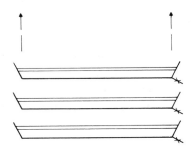

Fig. 9.6. Tray system for production of citric acid.

Despite the apparent crudity of the process, it is a reliable and economic synthesis of citric acid. Since the development of enclosed fermenters and acid resistant stainless steels, submerged fermentation has become possible. Production of toxic oxalic acid is suppressed, even at pH 5, by maintaining a low level of Fe in the medium; this is achieved by use of low Fe ingredients and addition of ferrocyanide as Fe-complexing agent. Although a more efficient process than the tray system, capital and running costs are higher, so the original system has survived in a few plants. Provided the fermentation is operated efficiently both processes are more economical than chemical synthesis or extraction from citrus fruits.

Production of lactic acid as an industrial chemical or an ingredient or additive in foodstuffs is carried out in enclosed fermenters using homofermentative thermotolerant or thermophilic lactic acid bacteria. A fermentation temperature of 45–50°C both accelerates the fermentation and suppresses contaminants; growth of saccharolytic clostridia would be a hazard at lower temperatures. Since streptococci or lactobacilli require no oxygen for growth the process is operated without aeration.

The nutrient requirements of streptococci and lactobacilli dictate a complex medium, but many possible ingredients, for example peptones, are unacceptable by their adverse effect on the extraction process. Glucose medium supplemented with yeast extract has been used successfully, as has whey, but molasses is avoided. Precipitation by chalk or lime, followed by treatment with sulphuric acid, yields a crude grade of lactic acid, adequate for some industrial applications. Further purification is achieved by distillation: lactic acid itself cannot be distilled, but a continuous process of esterification with methanol or ethanol, distillation and hydrolysis of the purified ester yields alcohol for recycling and a high grade lactic acid.

Citric acid and lactic acids are examples of industrial chemicals produced at least as economically by fermentation as by other methods. Little, if any, of the acetic produced by fermentation is used by the chemical industry, but acetic acid in the form of vinegar is widely used in the food industries: as a condiment in its own right and in preparation of pickles and sauces.

Synthetic acetic acid solution may be sold as a 'non-brewed condiment' but the name vinegar is reserved for the product of two successive fermentations. The source of alcohol may be cereals hydrolysed by malt (for malt vinegar) or grapes, raisins or grape juice (wine vinegar). Distilled spirit from these sources may also be acetified (spirit vinegar).

After the primary alcohol fermentation the clarified wash is oxidized by a mixture of *Acetobacter* and *Acetomonas* species. In the 'Orleans' process for production of wine vinegar, the film of acetic bacteria grows on the liquid surface in a 2/3-full cask of wine, the film being supported on a floating wooden grid. No inoculum is normally required initially; airborne acetic bacteria grow readily. As required, vinegar is drawn off and replaced, by the same volume of new wine, without disturbing the bacterial film.

Economical largescale production is by the 'German' process

(Fig. 9.7). A circular wooden vat of 6–8 m height and 3–4 m diameter, packed with wood shavings or twigs, is inoculated with the mixture of bacteria from an active fermenter. Although *Acetobacter* spp. oxidize ethanol to acetic acid and *Acetomonas* spp. continue the oxidation to CO_2, the exact nature of the bacterial mixture is unimportant since the process may be stopped at the desired stage of oxidation. Alcohol feed sprayed on the upper surface of the packing trickles over the microbial film on the twigs or shavings. One cycle is insufficient, but with recycling a satisfactory vinegar is produced. In some plants the process is continuous; in others batchwise. The legal minimum acetic acid content of vinegar, used as condiment, is 4%, but higher concentrations are required for pickling purposes and in practice the vinegar generator is operated to yield 8% acetic acid.

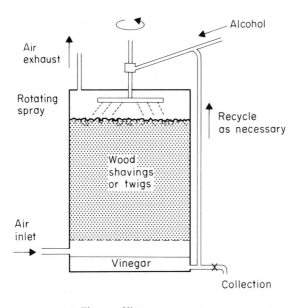

Fig. 9.7. Vinegar generator.

For some purposes the vinegar is distilled (distilled malt vinegar, distilled wine vinegar, etc.) and the clear liquid diluted to the required 4–8% acetic acid. Note that spirit vinegar is a different product, in which distillation occurred at the alcohol, not the vinegar, stage.

After fermentation, or distillation if applicable, vinegar is matured by storage, during which time esterification between acetic acid and residual ethanol, and traces of other alcohols, improves the flavour. Vinegar, with its trace content of esters, diacetyl, ethanol and higher alcohols, therefore, possesses a mixture of flavour components absent from the non-brewed condiment.

Single-cell protein and oil (SCP, SCO)

In various traditional oriental foods (for example koji, miso) the nutritional value of a foodstuff is improved by microbial, usually fungal, growth. Wheat or rice bran is a common starting material, and is rendered more digestible, more nutritious and more flavoursome by overgrowth of various fungi, but *Aspergillus oryzae* is a typical example.

This section is concerned with the recent developments in largescale culture of various micro-organisms to provide biomass of sufficiently high percentage of protein to be an acceptable foodstuff in its own right. Excess yeast of the brewing industry is sold for conversion into yeast tablets as a dietary supplement, or into yeast extract as a flavouring agent or an ingredient in various soup, gravy or snack-food preparations. Yeast grown specially as a foodstuff for humans, farm animals or domestic pets represents one of the most promising applications of the growth of microbial biomass as food. In particular, during the 1914–18 and 1939–45 world wars and their aftermath the aerobic growth of yeast cells produced food protein more rapidly than conventional agriculture. The often quoted comparison that a 0.5 t bullock synthesizes less than 0.5 kg of protein every 24 h, but 0.5 t of soya beans produces the equivalent of 40 kg protein every 24 h and 0.5 t of yeast generates 50 t in that time, illustrates well the main advantage of microbial SCP. Additionally, production of SCP requires little area and is largely independent of climate. On the other hand, the necessary technology is more complex, difficulties have been encountered with the high purine content of cells and indigestibility of cell walls and SCP is not as acceptable as animal protein in the diet.

Although many modern packaged foodstuffs are prepared with SCP, a more acceptable alternative was considered to be the large-scale production of SCP on cheap agricultural or mineral wastes, to provide a foodstuff for farm animals which would in turn become human food. Lipolytic yeasts grown on alkanes, and methanol utilizing bacteria and yeasts have been used for the production of SCP as animal food but many of these applications have been abandoned for reasons of economics, or safety and acceptability. Processes utilizing methanol appear to have the best prospects for the future and companies in the UK, USA and FRG have invested large sums in the design of fermenters and processing equipment for both bacterial and yeast systems. The ICI pressure cycle fermenter which is a combination of air lift and loop reactor with a working volume of over 100 m^3 has an annual output of approximately 70,000 tons SCP ('Pruteen'). The organism used is *Methylomonas methylotrophus* and the process operates as a continuous culture. Another bacterial process using a large tubular loop reactor has been designed by the Hoechst company in the FRG. In the USA Phillips Petroleum uses yeasts grown on methanol for the production of 'Provesteen'. The most effective organism is *Pichia pastoris* and the company claims cell densities of over 125 g cells (dry weight) per litre of fermentation medium. While at the present time the economics of SCP production from methanol are doubtful due to intensive competition from soya protein, etc., their potential for the future is not in doubt. In addition, and since methanol is likely to remain a relatively cheap fermentation feedstock for many years, many companies are looking to use these bacteria and yeasts for the expression of cloned gene products on a large scale. Hydrolysed wood wastes of the forestry and paper making industries also provided a suitable cheap nutrient source for SCP. *Saccharomyces* spp. utilize only hexoses and related disaccharides and require a growth factor supplement, but *Candida utilis* utilizes both pentose and hexose sugars, requires no addition of growth factors to the medium, can use either NH_4^+ or NO_3^- (or both) as N source and is nutritionally acceptable (after suitable treatment to reduce purine content) as food for both humans and animals. The production process is essentially the same as baker's yeast production; indeed in the past molasses was used as a nutrient for yeast SCP production but is now prohibitively expensive.

Another recent development in microbial SCP production is the use of fungal mycelium (for example *Fusarium*), which like soya beans, can be processed to the texture of animal meat. The *Fusarium* protein ('Mycoprotein') has passed all the toxicity trials required before being available for human consumption. Mushroom mycelium can be grown in fermentation vessels in the same way, usually for production of mushroom soup or pie mixes.

Human diet requires the amino acids lysine and methionine, which must therefore be present in satisfactory quantity in microbial or plant protein used as a complete substitute for animal products. Many SCP products are deficient in these amino acids, which therefore must be added to provide a satisfactory product. Many animal species similarly have an absolute requirement for these amino acids. Therefore the use of microbial SCP must be preceded by extensive testing for both toxic effects and nutritional value; often an expensive and prolonged process.

In addition to its content of cell proteins, the yeasts *Lipomyces* and *Rhodotorula* produce intracellular lipid, rich in the essential fatty acids, linoleic and linolenic acid. Under suitable conditions of aerobic culture over 50% of the cell mass of *Rhodosporidium sphaeroides* can be lipid. Subsequently, other, filamentous, fungi have been noted also to produce useful oils, and interest in the production of SCO for direct inclusion in diet, or as cooking oil, is developing rapidly at present. However an increased production by 'conventional' agriculture of various plant-seed oils (from soya beans, sunflower seeds, etc.) has restrained production so far of SCO. There is the potential, however, for the microbial production of specialized lipids such as linoleic acid as a replacement for 'evening primrose' oil.

Amino acids

Monosodium glutamate is the principal flavour component of soy sauce, and the conversion of soya bean extract to sauce is another example of traditional oriental fermentation technology.

Modern production of amino acids by fermentation was largely developed in Japan, and a number of amino acids including L-glutamate, L-lysine, L-aspartate and L-alanine can now be produced microbiologically in economic quantities. The principal advantage of the microbiological preparation is that only the L-isomer is formed. However, the availability of microbial amino acid racemases permits the previously poor yield of chemical synthesis to be boosted by conversion of the D-isomer.

The production and excretion of L-glutamic acid occurs in a wide variety of bacteria and fungi although the organism of choice is usually *Corynebacterium (Micrococcus) glutamicum*. Yields of up to 60 g g/L have been reported with a number of other genera including *Brevibacterium*, *Microbacterium* and *Arthrobacter* being able to excrete in excess of 30 g/L. All effective glutamic acid producers require biotin for growth, either lack or show only a low activity of α-keto glutarate dehydrogenase and have a high glutamic dehydrogenase activity. The production and excretion of glutamic acid depends critically on the permeability of the cytoplasmic membrane and this may be controlled by ensuring biotin deficiency in the later stages of growth, oleic acid deficiency (in suitable auxotrophs), and modification of the growth medium by addition of saturated fatty acids and of penicillin. A wide range of refined carbon sources may be used including fructose, maltose and xylose and complex sources such as cane and beet molasses. The use of refined substrates allows glutamic acid production to be controlled via modification of the biotin concentration. Since molasses has a high biotin content the use of these relatively inexpensive substrates require the addition of penicillin or fatty acids to ensure glutamic acid excretion.

Vitamins

Microbial cells, byproducts of various fermentations, have for many years provided a supply of mixed vitamins, especially as yeast extract. Most vitamins can be synthesized economically by chemical means, but pure riboflavin is produced competitively by microbiological synthesis. The most efficient production is by the plant pathogen *Eremothecium ashbyii,* a yeast-like fungus which produces up to 2 mg/ml of riboflavin extracellularly under suitable cultural conditions. An important feature is the association of vitamin production and excretion with growth of the organism, so, unusually among commercial fermentations, only a small inoculum, amounting to only 1% of the fermenter volume, is added to extend the logarithmic phase of growth. Extraction depends on the insolubility of the reduced form in water: riboflavin is precipitated from cell free culture medium by addition of a suitable reducing agent, and concentrated and redissolved by bubbling air through a slurry of the precipitate.

Another important vitamin is cyanocobalamin, vitamin B_{12}, the anti-pernicious anaemia factor. This compound is produced by various bacteria, but *Streptomyces griseus* is the normal producer of the compound. The quantities required for cure of pernicious anaemia can be adequately and most economically provided by occasionally altering the operational conditions of a streptomycin fermentation such that the vitamin, rather than the antibiotic, is the main product.

Although not strictly vitamins, carotenoids are precursors of vitamin A and are commercially produced as a dietary supplement, but more importantly as a natural colouring agent. Important applications are inclusion in poultry diets, to improve the colour of egg yolks, or diet of farmed salmon, to improve the colour of the flesh. Pink yeasts, for example *Rhodotorula,* useful as sources of SCP and SCO, provide also the carotene pigment for these purposes.

Flavourings

Traditionally, microbial growth has added desirable flavourings to a wide variety of foodstuffs and drinks: milk products, alcoholic beverages and soy sauce are three important examples. Various flavour compounds or flavour enhancers extracted from microbial cells are now widely used in the food industry. In addition to monosodium glutamate, probably the best-known flavour enhancers are the purine nucleotides.

Commercially, yeast is the most convenient source of these products, which are extracted as byproducts of the brewing, baker's yeast and SCP industries and used in soup, gravy and snack-food products. Alternatively, autolysed or smoked yeast may be used to simulate a wide variety of flavours. Although regulations for labelling of food products have to be scrupulously followed, it is possible to mimic many meat and vegetable flavours by suitably processed yeast extract and flavour enhancers.

The commercial production of 5′ IMP and 5′ GMP from the

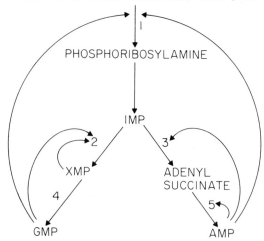

Fig. 9.8. Regulation of purine nucleotide biosynthesis in *Bacillus subtilis*. The activity of the first enzyme of this pathway (1, PRPP amidotransferase) is inhibited by AMP (strongly) and GMP (weakly). The enzymes of the GMP loop (2, IMP dehydrogenase and 4, GMP synthase) are inhibited by GMP and those of the AMP loop (3, adenyl succinate synthase and 5, adenyl succinate synthase and 5, adenyl succinate lyase) by AMP.

hydrolysis of yeast RNA has been carried out in Japan since 1961. The production of these compounds by direct fermentation usually employs cultures of *Bacillus subtilis*. These fermentations are good examples of the application of conventional genetic methods to industrial microbiology. The biosynthesis of purine nucleotides is illustrated in Fig. 9.8 which also shows the control exerted on enzymes of that pathway by feedback inhibition and repression. The production of IMP requires a mutation to adenine and perhaps guanine auxotrophy and the addition during the fermentation of sufficient concentrations of these compounds to allow growth but not feedback control. In addition, production organisms are usually mutated to resistance to toxic analogues of the purine bases such as 8-azaguanine and 6-methylthio purine. Such mutants lack part or all of the control of the early stages of purine biosynthesis by the end products.

Chapter 10. Fuels and chemicals

Three important applications of Biotechnology to energy supply operate at present: production of ethanol as liquid fuel, production of methane as fuel gas and the use of micro-organisms or microbial products in increasing the oil and gas yield of oilfields. Further applications of microbial or biochemical energy production are under active research, in particular the possibilities of microbial fuel cells.

Fuel alcohol

Awareness of the limited availability of nonrenewable fossil fuels, and the financial or political difficulties of various countries lacking indigenous oil reserves, have stimulated the production of ethanol fuel from renewable plant sources. Even with indigenous coal, its unsuitability as fuel for road or air transport requires an alternative source of liquid fuel. In most countries at present, conversion of coal to liquid fuels (and chemical feedstock) by chemical processes is more economical than generation of fuel by microbial fermentation, but this situation is unlikely to continue for long.

The production of fuel alcohol from sugar cane in Brazil illustrates the potential of microbially-generated fuel. Climate and land area suit the necessary scale of production, and sugar cane is an ideal renewable resource for fuel alcohol production. Factories built in areas of extensive cane cultivation obtain supplies at minimal transport cost. Juice pressed from the cane provides a satisfactory fermentation medium, and the residual cane debris (bagasse) is used as boiler fuel, supplying steam for stills, sterilization and local electricity production. Therefore there is no external fuel energy required by the operation, and the energy value of the ethanol is a net gain.

Conversely, production of fuel ethanol from maize has been suggested as a renewable energy resource, but the energy balance is unsatisfactory. The hydrolysis of starch to fermentable sugar, either enzymically or by acid hydrolysis, is an additional complication but the negligible fuel value of maize waste nullifies the energy balance of the system: the energy requirement for distillation is approximately equal to the energy value of the ethanol produced.

In special circumstances the fermentation may be economically or politically justified, despite the unfavourable energy balance. Apart from its importance in war conditions, fermentation of starchy or cellulosic wastes to fuel alcohol may be preferable as a waste-disposal system to energy-requiring but unproductive destruction of these wastes. Cellulose and hemicellulose are more appropriate sources of fuel ethanol than starch, which has potentially profitable use as a foodstuff. Unfortunately cellulose hydrolysis is more difficult than that of starch: despite extensive research, rapid cellulase activity has not yet been achieved, and the structure of cellulose complicates and retards acid hydrolysis. In addition to hexoses, hemicellulose products yield pentose sugars, not fermentable by *Saccharomyces* sp. and poorly fermentable, at present, by various yeasts of other genera.

Whatever the source of fermentable hexose, the fermentation is operated, from the highest practicable initial sugar concentration, with a high-yielding fast-fermenting strain capable of growth at 35°C, or higher if possible. Temperature tolerance is economical for two reasons: less cooling requirement, and less energy input later to reach distillation temperature. Distillation to the 96% ethanol:azeotropic mixture is normally followed by redistillation with continuously recycled benzene to provide anhydrous ethanol. Most applications of ethanol as motor fuel use a 10–20% addition to petroleum spirit which is satisfactory for standard spark-ignition engines. Anhydrous alcohol is then essential to prevent phase separation in fuel tanks. The 96% ethanol product may, rarely, be used alone as a fuel, but only in engines that have been suitably modified, particularly with respect to carburettor design.

Biogas

Methane, usually in combination with other combustible and non-combustible gases, is produced by anaerobic fermentation of sludge from industrial effluent treatment and provides a useful fuel for the operation of the treatment plant. The same principle, applied on a scale varying from wastes of a single dwelling to those of a substantial community, has been applied in recent years in African and Asian countries with very simple equipment (Fig. 10.1), to generation of fuel gas for cooking. This is considered further in Chapter 13.

Enhanced oil recovery

Contrary to popular belief, the oil reserves of a well are not in the form of a liquid pool; the oil is dispersed in the pores of the rock stratum. Normally less than 50% of the available oil is expelled by the pressure of gas trapped in the oil-bearing rock. Application

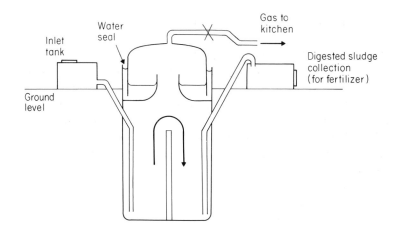

Fig. 10.1. Domestic scale biogas generator.

of additional pressure by water or recovered natural gas or, in landbased wells, CO_2, has been used to expel further oil, but successful results have also been obtained by biological methods.

Application of gas pressure has been achieved micro-biologically by injection of suitable barotolerant and thermo-tolerant bacteria along with fermentable sugar (molasses). Despite the apparent success, the risk of blockage of the porous structure of the rock by microbial growth is a serious disadvantage. Moreover, since the industry strives to prevent microbial con-tamination of oil wells, deliberate inoculation even with beneficial micro-organisms is viewed, generally, with acute concern.

A more acceptable recovery system relies on the high viscosity of microbial polymers to flush oil from the interstices of the oil-bearing rock. Various bacterial polysaccharides are available, chosen for stability at the temperatures and pressures encountered, but xanthan from *Xanthomonas campestris* is considered of great potential. If a sterile polysaccharide solution is injected, longterm damage to the well is minimized.

Surface active compounds in aqueous solution sweep oil from the rock more effectively than fresh or salt water. Although chemical detergents are often used for this purpose, surface active compounds of microbial origin are of interest; indeed the surface active nature of certain biopolymers (for example Emulsan) will contribute to their effectiveness.

Industrial chemicals

Ethanol and organic acids are important industrial chemicals; their production by fermentation was described earlier. The most economical method of production of many organic acids is by fermentation (Table 10.1) but direct production of industrial acetic

Table 10.1. Important chemicals produced by fermentation.

Ethanol
Acetone, butanol
Glycerol, mannitol and other polyols
Citric, lactic, gluconic and other acids
Amino acids
Antibiotics, vitamins, hormones

acid is not an economical process. Industrial ethanol was produced exclusively by fermentation until the 1930s, and the fermentation process competed successfully with the petrochemical process until the rapid expansion of that industry in the 1950s. The molasses byproduct of the sugar industry, a cheap and nutritious medium, was the preferred substrate.

After two decades of industrial ethanol production as, effectively, a means of waste disposal (using wastes of agriculture or the sugar-refining industry) the fermentation process has undergone a substantial revival, as a source of both fuel and chemical feedstock, since the late 1970s. Economics of ethanol production have already been discussed in the production of organic acids the cost of substrate, cost of extraction from fermentation broths, complexity of the molecule (including existence of isomers) and cost of purely chemical synthesis are economic factors which vary with each product. Often the preferred fermentation medium is based on agricultural products of unpredictable availability, and at a price varying with alternative demands for the product. Nevertheless the products listed in Table 10.1 are successfully produced by fermentation in competition with the present activities of the petrochemical industries, and can be expected to become more economical as that industry declines through exhaustion of oil reserves.

Acetone and butanol

Production of acetone and butanol by fermentation has suffered even greater decline than industrial ethanol. Introduced with some urgency in 1916 to meet the British demand for acetone for explosives during the 1914–18 war, it persisted with reasonable economic success until the early 1950s. Subsequently the acetone–butanol fermentation has been operated only as a method of waste disposal, on the premise that fermentation of agricultural wastes to a useful product was preferable to incineration; this applies especially in countries experiencing difficulty in obtaining oil and oil products. However, current concern over the diminishing oil reserves worldwide has stimulated a renewed interest in the process, and in particular with the development of strains capable of a more economic yield of solvents.

The acetone–butanol fermentation is important in the history of development of fermentation technology, as the first large-scale process requiring pure culture conditions. The alcohol and baker's yeast fermentations operated previously could be 'sterilized' sufficiently by steam at atmospheric pressure; the acetone–butanol vessels had to withstand steam sterilization pressures of at least 1 bar, and subsequently to operate under strictly anaerobic conditions. A further new complication was the sensitivity of the organisms to bacteriophage. In previous industrial fermentations, using yeasts, no such problem had been encountered.

The acetone–butanol fermentation as developed by Weizmann in 1916 used a strain of *Clostridium acetobutylicium*, but subsequently other strains of that species, or of other saccharolytic clostridia, were utilized. Although often used simply to avoid patent restrictions, many different species or strains of later fermentations did improve the yield of solvents.

Table 10.2. Products of acetone–butanol fermentation by *Clostridium acetobutylicum* (from glucose = 100).

CO_2	> 50
Butanol	> 20
Acetone	≈ 8
Acetic acid	≈ 4
Ethanol	≈ 3
i-Propanol	≈ 3
H_2	≈ 2

Relative amounts vary between strains, especially the acetone–butanol ratio. Not all compounds listed are produced by all strains of *C. acetobutylicum*.

Fermentation of sugars by *C. acetobutylicium* and other saccharolytic clostridia produces a wide range of compounds (Table 10.2). Amounts, and indeed the actual compounds, vary from strain to strain, but invariably the major product is CO_2. The gas (mainly CO_2+H_2) vigorously generated during fermentation provides sufficient mixing that mechanical stirring is unnecessary, and is used for pressurization and 'aeration' in the plant since air itself cannot be used with the strictly anaerobic bacteria. Such transfer of fermentation gas between vessels is a potentially serious source of phage infection, and in the early fermentations, many were completely inactivated in this way. Bubbling through chemical sterilant solutions was used for treatment of gas until in later years efficient filters, capable of retaining phage particles, were available. The alternative protection of using phage-resistant mutants was largely unsuccessful, since resistance was specific to only one type of phage, and resistant mutants gave poorer yields.

In its heyday the preferred substrate for the fermentation was molasses. Molasses became too expensive for production of low value solvents and, since the organisms are capable of hydrolysing starch, was replaced by various agricultural wastes. More recently, exploitation and improvement of the cellulolytic activity of clostridia has become more promising, to enable use of even cheaper waste products.

In the molasses process the medium, adjusted to 6% sugar and supplemented with N and P salts as required, was sterilized by at least 120°C and cooled to 35–40°C under a positive pressure of fermentation gas. Inoculum prepared over a series of cultures of increasing volume was added by gas pressure, and from the start of the fermentation the evolution of gas maintained anaerobic conditions and provided sufficient mixing for the bacterial culture. Initially the organisms produced acetic and butyric acids as products of fermentation, but as the pH fell below 6 the metabolism switched to production of the neutral solvents acetone and butanol, and much of the original acid was converted to solvents (Fig. 10.2). At the end of the fermentation, usually 40–48 h, the maximum tolerated yield of solvents (2%) had been reached and was recovered by distillation. Acetone, 'mixed solvents' comprising mainly ethanol and propan-2-ol, and the butanol/water azeotrope were collected, at increasing temperature levels on the still.

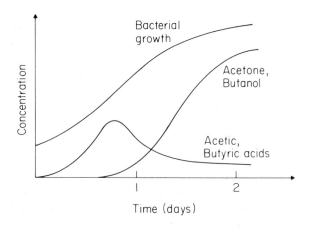

Fig. 10.2. Progress of acetone–butanol fermentation.

The fermentation gas, although the major product of the fermentation, was too impure to justify recovery for sale, but the microbial biomass provided a valuable source of riboflavin. Riboflavin is now produced more efficiently by other fermentations but the spent biomass of the acetone–butanol fermentation is of potential value as an animal-feed supplement. Although proteolytic clostridia are toxigenic, clostridia of the acetone and butanol fermentations are safe and acceptable as feed.

Polysaccharides

The complex structure of polysaccharides is difficult and expensive to synthesize chemically, but can be produced economically by microbial synthesis. A wide variety of polysaccharides of microbial origin is available (Table 10.3) and current research activity promises many additions to that list. Polysaccharide production is stimulated by excess of carbon source in the medium (and also excess N for N-containing heteropolysaccharides) but otherwise the fermentation is according to standard largescale culture procedures. Efficient stirring is essential for effective temperature control which becomes increasingly difficult as viscosity and cell mass increase during growth. Postfermentation removal of cells from polysaccharide is also more difficult because of the viscosity of the medium.

Table 10.3. Industrially or medically important polysaccharides produced by fermentation.

Dextran	Blood plasma substitute
Pullulan	Biodegradable packaging material
Xanthan	Gelling agent in foods, paints, drilling muds
Alginates	Gelling agents in foods

For many uses polymers of defined molecular size are required, and postfermentation processing is necessary to reduce as required to a specified range of molecular weight. Dextran, an α1-4 linked glucose polymer, is nonantigenic and is used as a plasma substitute. Dextran is synthesized by *Leuconostoc mesenteroides* growing on sucrose, but does not require input of energy: the energy of the glucose–fructose bond of sucrose is retained with the glucose unit after removal and metabolism of fructose. Therefore the free enzyme dextran sucrase is capable of performing the polymerization and provides an alternative method of dextran production, with the advantage that flow rate through columns of immobilized enzyme, originally produced by *L. mesenteroides*, can be adjusted to yield the optimum range of molecular size.

Chapter 11. Health care

The contribution to human and animal health is undoubtedly one of the most important benefits of modern biotechnology. In this chapter a small selection of examples is offered to illustrate the principles: antibiotics, vaccines and steroid hormones.

Penicillin

The lethal effect on Gram-positive pathogens of a diffusate from a *Penicillium* culture was first reported by Fleming in 1928. Although he realized the medical possibilities of penicillin, the stimulus for largescale production was the 1939–45 world war. Much of the technological expertise was already available, especially from the baker's yeast and acetone–butanol fermentations, but even so the rapid development of penicillin fermentation, extraction and purification on a largescale was an impressive achievement. Indeed, modern stirred aerated fermentation equipment developed simultaneously with the penicillin fermentation.

Penicillin is a secondary metabolite, produced after the period of rapid logarithmic growth of the organism. It is associated with nitrogen metabolism, and, although not realized at the time of the development work, the skeletons of the amino acids cysteine and valine are incorporated in the penicillin molecule (Fig. 11.1). However, these amino acids are not incorporated directly and penicillin production is depressed, not increased, by addition of these compounds to the medium.

In the development work of the penicillin fermentation in the early 1940s in Illinois, USA, it was fortunate that corn-steep liquor was chosen as the cheap nutrient for the fermentations. Corn-steep liquor is a byproduct of starch manufacture from maize: the preliminary soaking of the grain to extract unwanted nonstarchy compounds produces a liquor of high organic content. Concentration to 'corn-steep liquor solids' for use as an animal feed or culture medium ingredient is cheaper than effluent treatment. Phenylacetic acid is one component of corn-steep liquor, and supplied the benzyl group of the molecule of benzyl penicillin. Without phenylacetic acid as a 'steering agent', yield of penicillin, and activity, were substantially lower. The complexity of the composition of corn-steep liquor was another advantage: in the course of growth of the culture, glucose and other easily-metabolized

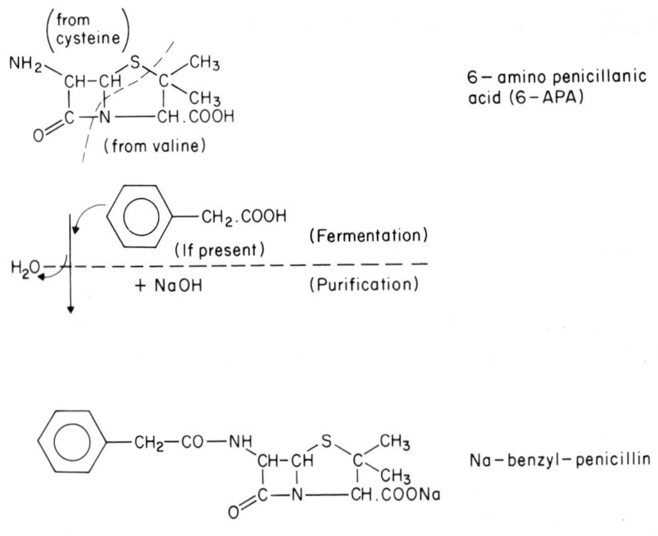

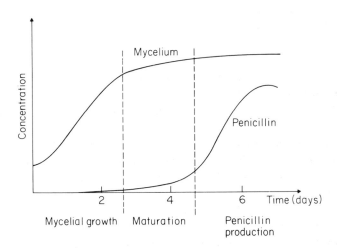

Fig. 11.1. Biosynthesis of penicillin (Na benzyl penicillin).

compounds provide the nutrient for the logarithmic phase of growth, and higher saccharides, organic acids and other slowly-metabolized compounds maintained activity later, during the phase of penicillin production (Fig. 11.2). Corn-steep liquor is no longer used as a nutrient; the nutrient requirements are met by a synthetic medium with sterile glucose solution fed as required for growth and penicillin production.

Fig. 11.2. Progress of penicillin fermentation.

Fleming's original strain, of *P. notatum,* proved unsuited for efficient largescale production, and after extensive search a better strain, of *P. chrysogenum,* was isolated in Illinois. Since then an extensive programme of genetic manipulation, by mutation, parasexual recombination and selection, has provided high-yielding strains. Early yields in USA of around 1000 units/ml (1 unit = 0.6 μg of sodium benzyl penicillin) were gradually increased to over 30,000 units/ml as a result of such development programmes.

Penicillin fermentations, and production of most other antibiotics, require 7–8 days for maximum yield (Fig. 11.2). Rapid and accurate biological assay was required to determine the optimum time of harvest, and a growth-inhibition assay with *Staphylococcus pyogenes* as test organism gives results within 4 h of sampling. Now, with modern computer-controlled fermenters the time of maximum yield can be predicted satisfactorily, and chemical assay of penicillin in the broth (by HPLC analysis) permits rapid confirmation of predictions.

In the early 1940s yields of penicillin were low (1000 units/ml, a good yield at that time, is equal to 600 mg/litre or 0.06%) compared with the production of the acetone–butanol or ethanol fermentations of the period. A specific and efficient extraction was essential, and was achieved by solvent extraction, originally in large steel separating funnels, but now by continuous countercurrent extraction.

Although the first to be developed, penicillin remains one of the best antibiotics available. Its high chemotherapeutic ratio (toxicity to target pathogen:toxicity to human) allows safe administration of high doses, and side reactions are rare. There were, however, a number of disadvantages. Penicillin was effective only against Gram-positive pathogens by inhibiting the synthesis of their specific cell wall structure. However, at the time of its introduction, Gram-positives caused by far the most serious infections. The structure of penicillin is destroyed by low pH, and very rapidly at 37°C. Except for treatment of mouth or throat infections, it could not be administered orally, being inactivated by stomach acid. Inactivation by penicillinase, which cleaved the 4-member lactam ring, was a hazard both during fermentation, when penicillinase-producing contaminants (for example *Bacillus* spp.) completely destroyed the yield, and in treatment, when penicillinase-producing pathogens were resistant. All of these disadvantages have been overcome with the development of 'semi synthetic' penicillins (Fig. 11.3).

The production of the various semisynthetic penicillins constitutes an interesting combination of microbiological and chemical syntheses. Penicillin is produced by fermentation as the starting material, and enzymically degraded to 6-aminopenicillanic acid (6-APA). Although 6-APA is produced by fermentation in synthetic medium lacking phenylacetic acid, the yield is low and the

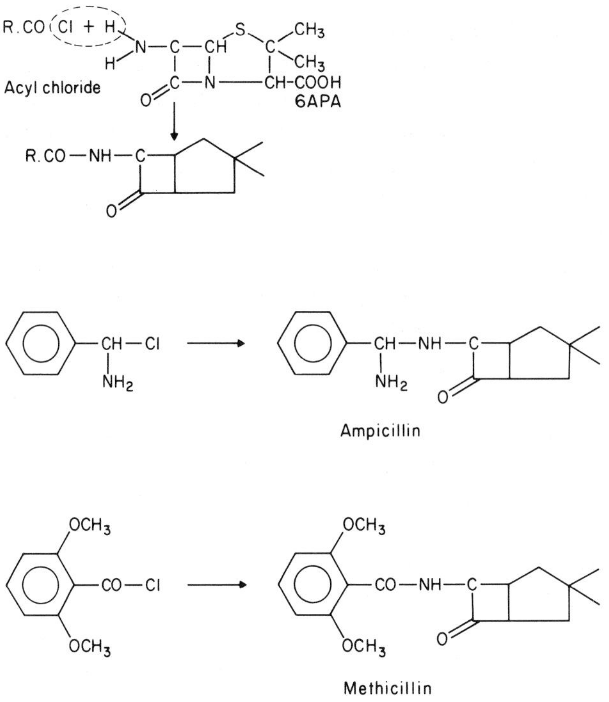

Fig. 11.3. Production of semisynthetic penicillins.

benzyl penicillin route is more efficient. Then, reaction with aminobenzyl chloride yields ampicillin (α-aminobenzyl penicillin) (Fig. 11.3) which reacts differently from benzyl penicillin, with activity against Gram-negative bacteria (but at the expense of weaker activity against Gram-positives) and acid resistance. Bulkier chemical groups, also reacted with 6-APA in the form of their chlorides, confer the additional protection of penicillinase resistance, presumably by blocking the access of the enzyme to the target lactam ring. Production of the range of semisynthetic broad spectrum antibiotics which can be taken orally, rather than given by injection as penicillin itself, has maintained the position of the penicillin fermentation as the most important antibiotic fermentation.

Other antibiotics

The realization of the importance of penicillin as a chemotherapeutic agent stimulated a deliberate search for other compounds; indeed, the search continues today. The next useful antibiotic to be discovered, streptomycin, is produced by the filamentous bacterium *Streptomyces griseus*. Being particularly effective against Gram-negative bacteria, its introduction in 1944 usefully complemented the activity of penicillin against Gram-positives. Streptomycin, structurally a trisaccharide (Fig. 11.4), blocks bacterial protein synthesis and impairs membrane function as the result of its binding to ribosomes. Although mammalian cells are largely protected by different membrane structure, streptomycin does show side effects on prolonged use, like many antibiotics developed after penicillin.

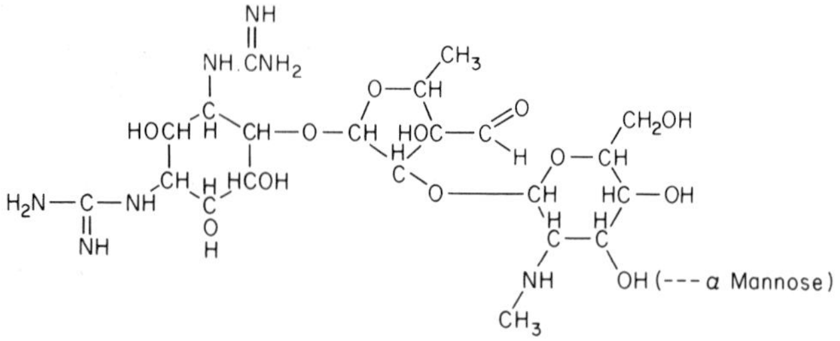

Fig. 11.4. Trisaccharide structure of streptomycin (mannose α-linked as shown to form streptomycin α-mannoside).

The streptomycin fermentation presented a number of additional problems: culture of the *Streptomyces* bacteria required a complex medium of neutral pH, and was sensitive to attack by bacteriophage (actinophage). Therefore more rigorous sterilization conditions were required for the medium, since the assistance of the acid pH of penicillin medium was no longer available. Although the air filters of penicillin fermenters were already of high efficiency, to remove spores of penicillinase-producing *Bacillus* spp., even greater efficiency was required for the streptomycin fermentation, to trap phage particles. Since streptomycetes are common soil bacteria, so too are their actinophages, and a further complication was the high aeration rate required for optimal yield of antibiotic. Fortunately, however, there is no enzyme corresponding to penicillinase, and the enzymic hydrolysis of streptomycin α-mannoside (Fig. 11.4) to streptomycin is encouraged; streptomycin itself is the more effective antibiotic. During the streptomycin fermentation, α-mannoside, cycloheximide (also known as actidione) and streptomycin itself are all produced simultaneously but by adjustment of culture conditions, including composition of the medium, production of any one of these compounds can be favoured. Cycloheximide is effective against fungi, but its toxicity to mammalian cells effectively limits its use to a useful antifungal component of laboratory culture media.

Postfermentation removal of *S. griseus* mycelium is difficult due to the smaller cell size than *Pencillium*. Addition of diatomite filter-aid allows effective clarification. Such addition was initially unwelcome, since in the early years of antibiotic technology, mycelial wastes were sold as animal feed supplement. Antibiotic residues in the waste suppressed intestinal bacterial flora and resulted in substantial improvement of the growth rates of calves, piglets, chickens and other farm animals. This practice has been largely abandoned as antisocially encouraging the development of antibiotic-resistant strains in nature.

The solvent extraction method for penicillin was not applicable to streptomycin. Ion-exchange resins provided the most effective method of separating antibiotic from components of the medium, and the various minor antibiotics from streptomycin itself.

Subsequent development of other antibiotics has followed the same principles. Initially discovered as a result of a screening programme using an arbitrarily chosen culture medium, the antibiotic is produced in sufficient yield for chemotherapeutic trials after determining the optimal conditions of aeration, agitation, pH, culture medium, etc., in small fermentation vessels. Finally the conditions, including optimum sterilization regime, for largescale fermenters, have to be determined, and the most efficient method of extraction. Few compounds ultimately reach this stage, but a continuing search is essential to counter the capacity of pathogens to develop resistance to established antibiotics.

Antibiotics are complex molecules, difficult and expensive to synthesize chemically (although this has been done in research laboratories to determine their structure) and alternative commercial production by exclusively chemical processes is unlikely. Chloramphenicol (Fig. 11.5), formerly produced by fermentation by *Streptomyces venezuelae* (as chloromycetin), is so far the only important antibiotic produced commercially by chemical synthesis. Therefore, in chloramphenicol we have an interesting link between the natural antibiotics and the entirely synthetic chemotherapeutic agents, for example sulphonamides.

Vaccines

Vaccine therapy was first introduced, although unwittingly, against smallpox, and was extended in use to a wide range of diseases by the work of the Pasteur and Koch schools of pioneer microbiologists. Ultimately prophylactic vaccination was practised against a wide range of viral and bacterial diseases, including intoxications, and immune sera were available for immediate emergency treatment of infections. Since the introduction of chemotherapeutic agents, including antibiotics, serum therapy against bacterial infections is no longer necessary. Otherwise, bacterial and viral vaccines, and antisera against highrisk viral diseases and bacterial toxins remain important products of health care biotechnology.

Bacterial cultures for use as vaccines are grown in stainless steel fermenters of 1000 litres or less, to facilitate handling of

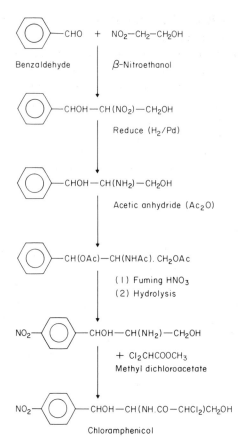

Fig. 11.5. Chemical synthesis of chloramphenicol (formerly produced as chloromycetin by *Streptomyces venezuelae*).

dangerous pathogens. An important deviation from normal fermenter practice is that after inoculation the internal pressure must never exceed atmospheric, to avoid risk of an infectious spray should any leak occur. Also, exhaust from the vessel must pass through a sterilizing filter, or incinerator, or both. Maintenance of stock cultures, and cultural conditions, are designed to maintain the virulence of the bacteria, or highest yield of toxin; only after harvest is the culture inactivated to provide dead vaccine or toxoid. In recent years the transfer of the virulence factor, where known, to *Escherichia coli* by 'genetic engineering' techniques has promised a safer production of bacterial vaccines; this may be the production method of the future.

Production of virus vaccines is, whenever possible, in tissue culture of appropriate cells, and this method has become virtually standard. Most fermenters designed for microbial culture are unsuitable for mammalian cells, which are damaged by the necessarily harsh conditions of aeration and agitation. With the introduction of unstirred airlift fermenters, largescale tissue culture has become available for medical and veterinary virus vaccine production. Unfortunately, cell lines capable of indefinite growth in tissue culture are of malignant origin, or suspected to have suffered malignant alteration, and are viewed with suspicion as a source of viral vaccines. Much of the present production makes use of freshly isolated cells, renewed as required.

Influenza, yellow fever and rabies vaccines were developed before tissue culture became available, and commercial production of the first two has continued to use fertile hen eggs. However, rabies vaccine production, formerly in animals, has been transferred to tissue culture and modern processing methods.

Steroid hormones

The production of steroid hormones is a combination of chemical and microbiological processes: in effect the reverse situation to the chemical modification of the microbiologically produced molecule to yield semisynthetic penicillins. Microbiological modification of molecules in an otherwise chemical reaction is of value when chemical modification produces two or more isomers, only one of which is active. Or the chemical reaction may have a mixture of compounds as the product, whereas the microbial product is one pure compound. Examples of enzymic or microbial transformations range from the simple conversion of glucose to sorbitol or gluconic acid, to the specific conversions of steroid and prostaglandin molecules.

Many hormones are steroids. Cortisone and hydrocortisone (Fig. 11.6) were known as adrenal hormones since 1930, but in 1950 it was shown that they suppressed the inflammation of rheumatoid arthritis. The suddenly increased demand could not

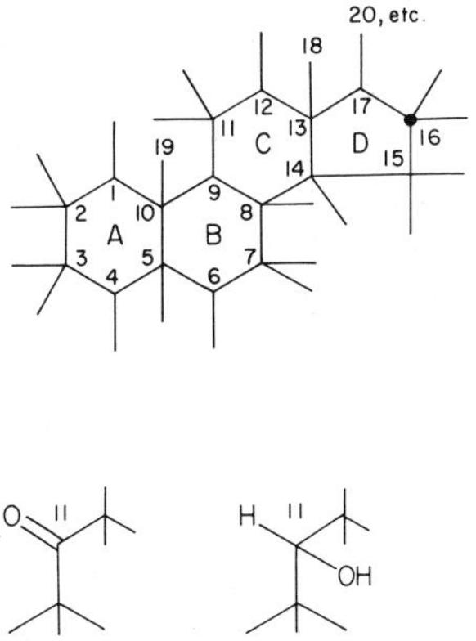

Fig. 11.6. General structure of sterols showing numbering of C atoms of the 4-ring skeleton. The detail underneath shows configuration of cortisone (left) and hydrocortisone (right) at C atom 11 of ring C.

be met by the existing source, the extraction from adrenal glands of slaughtered animals. Chemical conversion of relatively cheap sources of sterols, i.e. bile acids, lanosterol from wool, ergosterol from yeast or various plant sterol glycosides, was difficult, complicated and therefore expensive. The anti-inflammatory effect was shown to be associated with a keto or hydroxyl group at C-11 of the sterol nucleus (Fig. 11.6). The effect of —OH was stereospecific, and active and inactive isomers are formed in equal amount by chemical synthesis, effectively halving the yield. Worse, in order to add =O or —OH at a specific position (C-11), other positions on the molecule had to be temporarily blocked, and the protective acetyl groups had to be removed later: the number of reaction steps involved made semisynthetic hormones prohibitively expensive.

Therefore the realization that microbial enzymes mediated the addition or removal of specific groups, with no significant side-reactions, was a major advance in the production of steroid hormones, and the first commercial production, of cortisone, was introduced in 1952. Many different manipulations are performed now as a result of extensive search in a wide range of microorganisms for enzymic ability to modify sterol structure. The discovery of suitable organisms is purely by chance: there appears to be no way of predicting what transformation, if any, a species of micro-organism will perform. Many of the known reactions are best performed by a single strain of a species. Often different companies have discovered strains of very different species to perform a particular transformation: the species associated in Table 11.1 with specific reactions are certainly not the only possible organisms.

Table 11.1. Important transformation reactions at specific C atoms of the sterol skeleton.

1 Simple hydroxylation
2 Polyhydroxylation
3 Introduction of keto oxygen (=O)
4 Reduction of C=O to CHOH
5 Introduction or removal of double bond

The screening system for effective organisms, although tedious and labour intensive to perform, is in theory simple. As wide a range of micro-organisms as possible, usually of moulds, yeasts and bacteria, is tested in their normal growth media for an indication of the desired conversion. If any conversion occurs at all, the organism must have the necessary enzyme. Then follows the development work to determine the best fermentation conditions:

medium, pH, temperature, aeration, agitation, processing methods for the sterol. Is the original sterol toxic to the organisms? If so, but the product is nontoxic, add slowly. Obviously if the product is toxic, its toxicity limits the amount of original sterol added. Does the sterol product accumulate in the cells, or is it excreted into the medium? If in the cells, harvesting the cells offers a concentrated source of sterol, but with the disadvantage in the case of fungal conversions that the organism's own sterols complicate the extraction. Given the efficiency of modern continuous counter-current extraction equipment, it is often preferable to recover sterol from the medium.

The culture for the fermentation is developed through increasing size of seed fermentations in the normal way, but no sterol is included in the medium during these stages. Composition of the culture medium of seed stages is relatively unimportant; in practice that giving best cell growth is used. The medium of the production vessel must be as simple as possible, and restricted to the minimum for required growth, to facilitate subsequent extraction of the sterol product. Also, use of antifoam is undesirable: that, too, complicates sterol extraction. The vessel is inoculated and, normally, the culture grown through the logarithmic phase before sterol is added, i.e. to obtain maximum yield of cells, and therefore of the conversion enzyme, before the inhibitory sterol and solvent are introduced. Sterols are immiscible with water and a water-miscible solvent must be used, most commonly methanol, ethanol, acetone or propylene glycol. If the converting organism is capable of growing on oil as a carbon source, that can be used to dissolve the sterol, but will complicate subsequent extraction if added in excess. Often the amount of sterol added cannot exceed 2–5 g/litre, but with the high percentage conversion, often approaching 100%, and efficient extraction, this is acceptable. Water-immiscible solvents are required for extraction from clarified broth or harvested and macerated cells: methylene chloride, ethylene chloride or chloroform are often used.

When the required series of conversions (Table 11.1) cannot be performed by a single organism it may be possible to use a mixed culture. Noncompetitive reactions may be performed simultaneously, but many are competitive and are performed sequentially. Harvested sterol, redissolved in water-miscible solvent, is added to the production-scale culture of second and later conversions. Therefore by use of a suitable sequence of removals and additions of side groups, the cheaply obtained source sterol is converted to the required hormone. Most conversion steps are customarily achieved biologically, but removal of unwanted side-chains, or complete stripping of side-chains from the sterol nucleus, can if required be accomplished chemically as a first step.

The recent development of immobilized cell systems has substantially improved the efficiency of sterol conversions. Cells trapped in alginate, or other immobilizing agent, are held in a column through which sterol solution is passed. Commonly the sterol is dissolved in water-soluble solvent for passage through the column, and extracted later. Obviously the immobilized system is applied only when the sterol yield is entirely in the medium, but since no growth is expected in the columns the simpler composition of the medium benefits extraction. Satisfactory conversions have also been achieved by feeding sterol through the column in the water-immiscible phase, with obvious advantage to subsequent extraction.

Chapter 12. Agriculture

Initial interest in the agriculture market from biotechnology companies focused on long term, supposedly lucrative goals such as genetically engineering nitrogen fixation genes into cereal plants. Activity is now moving into a more product oriented approach and crop strain improvement, increased herbicide resistance and biological control of insect pests and plant diseases are all popular. Animal health care centres on the prevention of disease through the development of improved vaccines through genetic engineering and the design of synthetic peptides.

Animal vaccines

Diseases caused by viruses, bacteria and parasites are responsible for severe depletion of animal stocks worldwide and particularly in third world countries (Table 12.1). Treatment of disease with antibiotics is often possible in developed countries but expensive and economically impossible in poorer countries. Moreover young stock, which once weaned have little resistance to infections, can die within a short time of contracting disease even if

Table 12.1. Some major diseases of large animals.

Disease	Animals affected	Symptoms	Prevention	Treatment
Viruses				
Border disease	Sheep	Abortion		None
Foot and mouth disease	Cattle sheep pigs	Loss of condition, death	Slaughter	None
Enzootic abortion	Sheep	Abortion	Vaccination	None
Bacteria				
Anthrax	Cattle pigs	Fever, death	Vaccination	Antibiotics
Clostridal diseases: blackleg braxy lamb dysentry struck tetanus	Horses cattle sheep pigs	Death Indigestion and death Death Spasm and death	Vaccination	Antibiotics
Mastitis	Cattle	Swollen udder, fever	Good husbandry	Antibiotics
Tuberculosis	Cattle	Death	Slaughter	None

correct treatment is administered. Vaccination is therefore an extremely effective approach to protecting animal populations from disease. Not only is it the only available procedure for virus infections, it can provide longlasting immunity and can limit the spread of pathogenic organisms in a normally susceptible population. Vaccines are currently available for a wide range of pathogenic organisms for cattle, sheep, pigs, poultry and fish.

There are problems with many of the existing vaccines however. In the case of live attenuated strains, mutation to increased virulence or loss of antigenicity is always a problem and preparations need rigorous testing. Similarly, inactivated vaccines must be carefully screened to ensure that live, virulent organisms do not evade the inactivation process. In general, both attenuated and inactivated vaccines must be carefully shipped and stored to prevent loss of potency, often a problem in tropical regions. Finally, some micro-organisms are so poorly antigenic that it has not been possible to manufacture effective vaccines by traditional procedures. Thus the possibilities for modern molecular biology in vaccine development both for the improvement of existing vaccines and the introduction of new ones are enormous. It has been estimated that new vaccines for livestock and poultry have a potential worldwide market value of $4000 million by 1995. Two examples will be described to outline the major approaches.

Foot and mouth disease (FMD) of cloven-hoofed animals is caused by a picornavirus, a small single stranded RNA virus. Vaccines, prepared by inactivation of virulent virus grown in baby hamster kidney cells have successfully controlled the disease for several decades and it is eliminated from some countries. Current worldwide production of FMD vaccine approaches 10^9 individual doses, but the material has several disadvantages. There are seven principal serotypes and many more subtypes which complicates effective protection. As mentioned above, virulent virus may survive inactivation and the vaccine requires refrigeration during transport and storage. FMD virus has therefore been chosen as a suitable system for engineering genetically a more effective and cheaper vaccine. The major virus antigenic determinants are located within the protein VP1. The cDNA route was used by several groups to clone the VP1 gene and efficient expression in *E. coli* has been obtained using the *trp* promoter to direct the synthesis of an insoluble hybrid protein that comprised 17% of the total cell protein. VP1 proteins produced in *E. coli* induce virus-neutralizing antibodies in cattle and swine and can protect animals against virus challenge, but the immunizing activity is low compared to the intact virus particle and at present these preparations cannot compete economically with the traditionally produced vaccines.

An alternative route to improved vaccine development is to

synthesize synthetically a small peptide that mimics the antigenic determinant of the protein. Of the many antigenic sites in a protein those that are located on the surface in a highly exposed situation are usually the most immunogenic. Such sites are generally hydrophilic, containing arginine, lysine and aspartate residues. Moreover, proline is often found where a polypeptide chain makes a sharp bend and is associated with antigenicity. In the case of FMD virus, the sequence of the 213 residues of VP1 is known and the region between amino acids 141 and 160 conform to the above criteria. Moreover, in different serotypes, the composition of this region varies suggesting that this is the antigenic determinant. Synthesis of the peptide spanning the region 141–160 and coupling it to a carrier molecule provided rabbits with neutralizing antibody to the virus when inoculated with the synthetic vaccine. Furthermore cattle immunized with the peptide secrete antibody against FMD virus and it may be that this will be the most effective approach to the development of new vaccines.

Transgenic animals

Conventional animal husbandry has resulted in breeds that are suited to particular climates and/or purposes. The possibility of transferring genetic material from one breed to another would speed this process enormously and more dramatically, could enable genetic transfer between species.

The manipulation of transgenic mice is now an almost routine technique but recently it has been extended to large mammals namely rabbits, pigs and sheep. Such animals present their own problems. For example a single superovulated mouse might produce 40 eggs but a ewe gives just 3 or 4. Mice can tolerate up to 20 reimplanted embryos but sheep or cattle will only give birth to 1 or 2. The technology must also be modified. For example to obtain high transformation frequencies the DNA should be injected into the nucleus but in some instances the cytoplasm of the egg is too opaque to achieve this. Interference microscopy has helped in this connection and for pig embryos it was found that centrifugation clarified the cytoplasm.

It is therefore encouraging that the mouse metallothionein–human growth hormone gene has been integrated into rabbits (12–8% success rate) and pigs (11%) almost as successfully as mice (27%). Moreover some 10% of the infected embryos survived in their foster mothers. Expression of the gene was observed in some of the transgenic pigs and rabbits.

What then are the aims and benefits of this work? The obvious answer is to increase the size of animals by incorporating additional copies of growth hormone genes. Such schemes have been initially successful in a variety of animals from salmon to cows but in other instances, such as poultry, increasing growth hormone levels has led to a reduction in growth rates. Obviously a more detailed knowledge of the physiology of these animals is needed. Other possibilities include increased lactation in cows, improved reproductive performance by altering hormone balances and the introduction of keratin genes to alter the properties of hide and wool.

Micropropagation of crop plants

One of the major uses of micropropagation techniques is the establishment of disease-free stocks. Disease from viruses, viroids, fungi, bacteria and nematodes are responsible for reduction in yield and impairment of crop quality. With bacterial, fungal and nematode infections, explant material can be removed from a noninfected part of the plant and micropropagated. In the case of viral diseases apical meristem material is used since these are the only cells unlikely to be infected. Through the clonal propagation of this material, disease-free stock can be rapidly accumulated. This approach has been particularly successful for the production of virus-free potato stocks, many ornamental, flowering plants and fruit crops.

For many crop plants, seed production is difficult and expensive, particularly where hybrid seed must be generated. Clonal propagation offers the possibility of largescale multiplication and maintenance of breeding lines but a limiting factor is the high degree of genetic conservation that is required. Nevertheless, the production of bulked parent clones for the production of uniform seed of F′ hybrid lines in crops like cauliflower have been valuable.

Mass propagation of crop and forest plants is an important application of micropropagation techniques. Somatic embryogenesis involves the development of embryos from somatic cells in culture. The embryos bear both root and shoot apices and thousands can be grown in one flask. Although not all plants are amenable to this technology, several crop plants including alfalfa, cotton, citrus, carrot, oil palm and pearl millet can be handled in this way although few systems show immediate promise. Several largescale clonal propagation programmes are now underway in particular the planting of cloned oil palm has been taking place in Malaysia since 1977.

Genetic improvement of crop plants

Although genetic variation amongst cultured plant cells and the plants derived therefrom can be problematical when preparing material for propagation, it can be harnessed as a useful tool for introducing variability into cultivated plants. The chromosomal changes often include polyploidy and structural modifications to the chromosomes. Chromosome rearrangements and single gene mutations have also been detected, albeit at lower frequency on plants regenerated from cultured cells. By using appropriate selection procedures it should be possible to recover mutations for inclusion in crop development programmes. This approach has been particularly useful for the generation of variants resistant to disease. Sugar cane strains with resistance to Fiji disease, downy mildew and eyespot disease have been isolated, potatoes with resistance to blight, and yield improvements and resistance to tobacco mosaic virus in tobacco have been reported. New hybrids with disease resistance and other desirable traits might also be generated by protoplast fusion between cultivars and species, this technique has been proved successful with tobacco but reproducible methods for use with crop plants are not yet available.

Genetically engineering resistance to disease by recombinant DNA methods has been hampered by insufficient knowledge of the bases of pathogenicity and resistance mechanisms. However, one area in which progress is being made is in engineering herbicide resistance. Although herbicides are usually chosen and developed to be harmless against the crops they are designed to protect, in several instances special tolerance to a herbicide could be very important. Such an example is the Du Pont herbicide Glean (chlorsulfuron), an effective, broad spectrum herbicide used on wheat and barley. It kills most other plants and residues in the soil can damage crops planted after the cereal. Tolerance to Glean in other crops would therefore expand the utility of this product. As a first stage, *Salmonella* strains resistant to Glean were isolated and the phenotype was mapped to the *ilvG* gene which encodes the first enzyme in the pathway leading to isoleucine and valine. The gene was transferred to tobacco cells in culture and the regenerated plants were resistant to the herbicide. Similarly genes conferring tolerance to glyphosphate have been developed in enteric bacteria and transferred to tobacco.

Genetic improvement of plants also encompasses modification of the nutritive value of crops. A typical example involves the genetic manipulation of maize (corn) which is used for human consumption and fed to poultry and pigs. Like most other cereals, it is deficient in the essential amino acid lysine. Currently lysine is added to animal feed based on maize as the pure amino acid derived by fermentation or in soya protein, but both remedies are expensive. Current research has identified the major storage protein (zein) genes in corn and the next stage is to modify the genes such that the lysine content will be increased, although it is unlikely that this will be achieved in the near future.

An often-quoted example of genetic engineering to improve crop yields concerns nitrogen fixation in plants. Most crop plants do not fix atmospheric nitrogen and increases in yield rely on the application of nitrogenous fertilizer. Strains of rice, wheat or corn that use atmospheric nitrogen would have obvious benefits. Nitrogen fixation is largely the preserve of rhizobia in symbiotic combination with leguminous plants. Indeed a great deal of effort is currently devoted to understanding this complex relationship and engineering rhizobia for increased nitrogen fixation and nodulation. It seems unlikely however, that a symbiotic interaction between nonleguminous plants and a rhizobium can be achieved. The alternative approach is therefore to place genes for nitrogen fixation directly into the plant of choice. This may seem simple superficially but a brief examination of the project should outline the complexity of the problem and the still somewhat primitive stage of plant molecular biology.

The most fully understood, free-living nitrogen-fixing bacterium is *Klebsiella pneumoniae*. The 17 nitrogen-fixing (*nif*) genes are located in 8 operons of which one encodes the three subunits of the nitrogenase enzyme. Control of transcription is complex; briefly the products of *ntrA* and *ntrC* activate transcription of *nifA* the product of which together with *ntrA* product activates transcription of the other genes.

Somehow these 17 genes must be introduced into the plant and expressed. At present the Ti plasmid is the only reasonably advanced vector and it is not suitable for monocotyledonous plants such as cereals. Assuming transformation of the gene cluster into the plant, expression must be engineered. As pointed out by Merrick and Dixon (1984) the chloroplast offers a more suitable site for the *nif* genes than the nucleus. Mechanisms controlling expression of genes in the nuclei are different from those in bacteria, and although it is feasible to provide single prokaryotic genes with the requisite sequences for transcription and translation by using Ti plasmid T-DNA promoters for the 15 genes of the *nif* cluster, this would require extensive manipulation. The chloroplast on the other hand is a probable evolutionary descendant of a prokaryote and retains a bacterial-type genetic system. It is therefore likely that the chloroplast might express *nif* genes with a minimum of manipulation, particularly since *E. coli* will express the chloroplast ribulose bisphosphate carboxylase gene. A scheme for the proposed expression of *nif* genes in plant chloroplast is shown in Fig. 12.1 but as will be appreciated, strains of nitrogen-fixing corn are still a long way off.

Microbial control of insects

With increasing worries concerning the environmental implica-

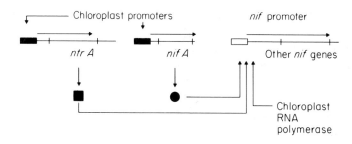

Fig. 12.1. Proposed manipulation of *nif* regulatory genes to engineer *nif* gene expression in a plant chloroplast. The *ntrA* and *nifA* genes could be placed under the control of chloroplast promoters to ensure optimal expression. Assuming the *ntr* and *nifA* gene products could interact effectively with chloroplast RNA polymerase, expression of *nif* genes would come under the control of the chloroplast promoters fused to *utrA* and *nifA*.

tions of chemical pesticides, attention is turning to the simple yet elegant idea of controlling insect pests of plants and animals with infectious disease agents. Such microbial control agents include viruses, bacteria and fungi and have been commercially available since the 1940s but their sector of the pesticide market has remained small. Now, experts suggest that biological control agents could invade existing markets and develop new markets such that by the next century they might comprise half the world insecticide market. What then are the advantages of biological control agents?

Firstly they are 'safe' compared with their generally more toxic chemical counterparts. Secondly, they can be persistent and give lasting control, although they do not accumulate in food chains. Moreover, although they have not been used as extensively as chemical agents, there is evidence that resistance does not develop so rapidly, if at all, in the insect population. Resistance to a chemical may require a single mutation, the pathogenesis of disease tends to be more broadly based. Thirdly, microbial control agents are often compatible with other insecticides, including chemicals, and can be used in combination with them. There are some drawbacks however. Microbial control agents tend to be slow-acting and are not readily accepted by the farmer who prefers to see a rapid and dramatic effect from spraying. Secondly, microbials are often too specific, killing only one insect type and only during a certain developmental stage. This can sometimes be an advantage since the ecosystem is not overly disturbed but more often it is disadvantageous. Finally there are problems in the production of the micro-organisms, maintaining viability and infectivity and formulating effective spraying programmes. Nevertheless, these latter aspects will surely be redressed considering the current impetus for research.

There are essentially two approaches to microbial control of insects. Micro-organisms can be used in the same way as a chemical agent and sprayed repeatedly as and when necessary. These micro-organisms do not usually persist in the field. Included in this category is the most successful of all microbial control agents, *Bacillus thuringiensis* (known as B.t.). As this bacterium sporulates it synthesizes an intracellular protein crystal, the parasporal body. Although the bacterium is weakly pathogenic to various larvae of the Lepidoptera (butterflies and moths) the crystal protein is highly toxic. The insect larvae ingest the toxin and spores when feeding on contaminated (sprayed) vegetation; the toxin causes gut paralysis and the bacterium is then able to invade the weakened host causing a lethal septicaemia.

Commercial preparations of B.t. are usually wettable powders or liquid suspensions of spores and crystal protein. They are effective on some 200 crops against 55 pest species especially inchworms and gypsy moth caterpillars. The main advantages are: (1) safety, B.t. has no effect on vertebrates including man; (2) relatively rapid control similar to a chemical pesticide; and (3) no resistance has been reported. On the other hand, the spores and to a lesser extent the crystals are inactivated by prolonged exposure to sunlight and the B.t. insecticides are not cheap; they can cost 20–25% more than equivalent chemicals. However, the toxin gene has now been cloned and expressed in *E. coli* and improvements in yield and spectrum of activity will no doubt ensue. This will enhance the cost price efficiency of B.t. insecticides.

Several serotypes of B.t. of which var. *israelensis* was the first to be reported are toxic to the larvae of mosquitoes and blackfly, again through the synthesis of a crystal protein but the molecule differs from the Lepidopterous toxin. Formulations are commercially available for the control of these important vectors of human and animal disease. Similarly, some strains of *Bacillus sphaericus* are toxic to mosquito larvae and since these bacteria are reported to replicate in the field for up to three months they could give longer lasting control.

The second approach to microbial control of insects involves introducing a disease into an area with the aim of accomplishing proliferation and spread of the agent such that a persistent and endemic infection ensues. One successful example involved the 'natural' introduction of insect pathogenic viruses into North America to control forest pests. There are six main groups of insect viruses of which the Baculoviruses in particular, and Reoviruses are so different from vertebrate viruses that they have potential as safe control agents. The European spruce sawfly was a serious threat to Canadian forests earlier this century following its accidental import from Europe and the lack of indigenous parasites in North America. A variety of natural parasites were imported from Europe to no avail but a viral disease emerged and spread presumably following accidental introduction along with the parasites. The sawfly population decreased and has not since been a serious problem. Other viral agents include strains of Baculovirus for control of cotton bollworm, tobacco budworm and Douglas fir tussock moth. Nevertheless, the future for viral insecticides is uncertain although the potential seems great.

Bacillus popilliae, a highly fastidious aerobic, endospore forming bacterium causes milky diseases of some beetles in particular the Japanese beetle an important pasture pest. Unlike *B. thuringiensis* these bacteria are inherently pathogenic. When the spores are ingested by the larvae, they germinate in the gut and the bacteria grow causing the transparent larvae to adopt a 'milky' appearance. The larvae die and the spores from the bacteria are released and ingested by foraging larvae. Thus a persistent, control of the beetle is obtained. One of the major drawbacks was the inability to produce spores of this bacterium *in vitro* and thus the expense of preparing the insecticide *in vivo*. However, recent research has focused on this problem and the requirements for largescale cultivation of the bacterium and its spores have been defined. Improvements in the host range (which is rather narrow) would now be desirable.

Fungi have several advantages as insect control agents. They tend to have broad host ranges, are infective upon contact and do not have to be ingested. On the other hand they are more sensitive to environmental conditions like temperature and moisture and to chemical fungicides. Most insect-pathogenic fungi are deuteromycetes or 'fungi imperfecti' and include *Beauveria bassiana* and *Verticillium lecanii* which have been used to control Colorado beetle of potatoes and aphids respectively. *B. bassiana* is also being developed to protect citrus crops from mites.

In conclusion, the success of microbial insecticides will depend on the ability of biotechnology to provide improved and effective strains of virus, bacteria and fungi. Just as importantly however, the farmer must be persuaded to change from the familiar efficient chemicals to the biological alternatives and educated in their proper use and application. The question of safety has yet to be fully answered and regulatory approval may be problematical. Nevertheless the potential is great and combinations of biological and chemical control agents will probably be used in the future.

Products from plant cell cultures

There are essentially three commercial applications from the bulk cultivation of plant cells in fermenters. Although it seems unlikely that plant cells will ever compete with microbial systems for the production of biomass, largely because their growth rates are too slow (doubling time 25–100 h), nevertheless yields may be high (5–20 g dry wt/litre) and the process economically viable for specialized cases such as tobacco. However, the principal applications are currently the production of various, high cost, low bulk materials and biotransformations.

It seems likely that any product currently extracted from plants could be derived from *in vitro* cultures. The range of such materials is diverse and includes pharmaceuticals, insecticides, flavourings and perfumes (Table 12.2). The bulk commodities such as latex, enzymes, waxes and oils will not be dealt with here since it seems unlikely that they will be derived *in vitro*, at least in the near future. Plant derived medicinal compounds are still used widely despite the development of synthetic drugs and comprise some 25% of prescribed drugs. They include steroids, analgesics, cardiovascular and antimalarial agents (Table 12.2) and particularly where large amounts of plant material have to be processed to obtain a small amount of drug, the benefits of high-yielding cultures growing in a fermenter are obvious. In addition to these established compounds, novel drugs, flavouring agents, insecticides and other fine chemicals are waiting to be discovered and it is not surprising that screening programmes are increasingly being directed towards plant products.

Table 12.2. Natural products from plants and their associated industries (after Fowler 1983).

Industry	Product	Plant species	Industrial uses
Pharmaceuticals	Codeine (alkaloid)	*Papaver somniferum*	Analgesic
	Diosgenin (steroid)	*Dioscorea deltoidea*	Antifertility agents
	Quinine (alkaloid)	*Cinchona ledgeriana*	Antimalarial
	Digoxin (cardiac glycoside)	*Digitalis lanata*	Cardiatonic
	Scopalamine (alkaloid)	*Datura stramonium*	Antihypersensitive
	Vincristine (alkaloid)	*Catharanthus roseus*	Antileukaemic
Agrochemicals	Pyrethrin	*Crysanthemum cineraviaefolium*	Insecticide
Food and drink	Quinine (alkaloid)	*Cinchona ledgeriana*	Bittering agent
	Thaumatin (chalione)	*Thaumatococcus danielli*	Non-nutritive sweetner
Cosmetics	Jasmine	*Jasminium* sp.	Perfume

Economic product yields remain a problem, particularly for secondary metabolites for which some differentiation is required before gene expression is optimal. But it is likely that the complexities of plant gene expression will be increasingly scrutinized by the molecular biologist and stable cell lines capable of high levels of regulated expression of pertinent genes will undoubtedly be constructed. In the meantime it is encouraging that yields of some products in cell culture approach those found in whole plant tissue (Table 12.3) and some processes such as the manufacture of antitumour alkaloids by cells of *Catharanthus roseus* are now reaching pilot plant scale. Others including phosphodiesterase and ubiquinone 10 from tobacco cells and shikonin from *Hithio spermium* sp. are being produced on a large industrial scale.

Table 12.3. Yields of some natural products in cell cultures and their equivalent amounts in whole plant tissue. (From Mantell *et al.*, 1985.)

		Yield	
Natural product	*Species*	*Cell culture*	*Whole plant*
Anthraquinones	*Morinda citrifolia*	900 nmol g^{-1} dry wt	Root, 110 nmol g^{-1} dry wt
Anthraquinones	*Cassia tora*	0.334% fresh wt	0.209% seed, dry wt
Ajmalicine and serpentine	*Catharanthus roseus*	1.3% dry wt	0.26% dry wt
Diosgenin	*Dioscorea deltoidea*	26 mg g^{-1} dry wt	20 mg g^{-1} dry wt tuber
Ginseng saponins	*Panax ginseng*	0.38% fresh wt	0.3–3.3% fresh wt
Nicotine	*Nicotiana tabacum*	3.4% dry wt	2–5% dry wt
Thebaine	*Papaver bracteatum*	130 mg g^{-1} dry wt	1400 mg g^{-1} dry wt leaf and 3000 mg g^{-1} dry wt root
Ubiquinone	*Nicotiana tabacum*	0.5 mg g^{-1} dry wt	16 mg g^{-1} dry wt leaf

Cultivated plant cells can be used to effect transformations of commercially important chemicals in the same way as strains of bacteria and fungi are used in steroid transformations. A variety of transformations have been achieved using plant cells including

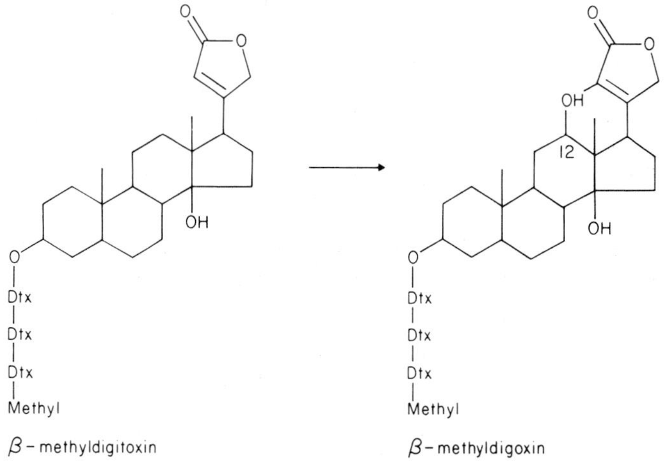

β–methyldigitoxin   *β*–methyldigoxin

Fig. 12.2. Digitoxin can be methylated by chemical means to β-methyldigitoxin which is transformed by selected cell lines of *Digitalis lanata* very rapidly and efficiently to β-methyldigoxin. Dtx, digitoxose molecule.

hydrogenations, hydroxylations, oxidations, etc. The most successful biotransformation in a commercial sense is the production of the cardiac glycoside β-methyldigoxin from β-methyldigitoxin using cell cultures of *Digitalis lanata*. Both digoxin and digitoxin are extracted from *D. lanata* for the treatment of heart disease but digoxin is used more frequently than digitoxin. Since the latter is produced during the isolation of digoxin, it accumulates. In one process, β-methyldigitoxin is prepared from digitoxin by chemical methylation and *Digitalis* cell cultures grown in 200 litre downdraft airlift fermenters are used to effect the 12 β-hydroxylation of this molecule to form β-methyldigoxin (Fig. 12.2). This biotransformation has also been achieved with immobilized cells of *D. lanata* which, with the longevity of the immobilized cells should make the process even more economical.

Chapter 13. Waste treatment systems and biodegradation

In terms of volumetric throughput, the treatment of sewerage and waste water from domestic and industrial sources is the largest biotechnological industry. In the UK the fermentation capacity of the waste water industry is estimated at approximately 3 thousand million cubic metres (or some 20 times that of the brewing industry) and most industrialized countries follow a similar pattern.

The treatment of domestic and industrial wastes has a long history and in many industrialized countries sewerage systems and treatment plants were installed initially some 100 years ago. These were built as the epidemiology of many infectious diseases which were the scourge of crowded urban populations became known. Over the last 20 years, however, there has been a steady raising of the standards required by river authorities before wastes may be discharged into water courses with severe financial penalties for failure to conform. The strength of liquid effluents is measured by the concentration of suspended solids (SS) together with the concentration of organic material which can be oxidized chemically on boiling with potassium dichromate and concentrated sulphuric acid (the chemical oxygen demand or COD). The COD limit of effluent to be discharged into a natural water course is usually of the order 10–20 mg per litre. A typical brewery bulked effluent might have a SS of 240 mg per litre and a COD of 1800 mg per litre. The necessity to treat such wastes arises from a need to reduce the COD and the associated biological oxygen demand (BOD) of the effluent stream. Failure to do this would result in dire consequences in a water course due to high microbial activity and a concommitant lowering of the oxygen content of the water. In addition some effluent streams might have extremes of pH or may contain toxic chemicals which require special attention.

Types of treatment systems

Many types of wastes are suitable for biological treatment and some may even yield useful products. For example, wastes from intensive cattle rearing may be fermented anaerobically to produce methane as a fuel and SCP for animal feed purposes may be produced from the waste streams of food processing plants and paper and pulp mills. Other wastes may be more problematical, for example oily water discharges from the petrochemical indus-

142

Chapter 13
Waste
treatment
systems and
biodegrad-
ation

tries and distillery wastes with a high COD and a high concentration of salts including Cu and SO require special treatment, the latter often being discharged onto waste land. Wastes which are especially toxic or of very high COD are often treated in specialized plants within the chemical works producing them while wastes of lower toxicity or with only a moderate COD are dealt with in municipal treatment plants. Some chemicals may resist biodegradation and this may be especially true of those of man-made origin (xenobiotics).

Dry disposal of wastes involves compacting material into layers alternating with soil and finally covering with soil. Such landfills are obviously limited by the availability of suitable sites but continue to be used extensively by municipal authorities. The decomposition occurs slowly and anaerobically with methane being the expected product. In some instances this methane may be of nuisance value with the possibility of combustion occurring or (potentially) may be tapped via boreholes and used as a fuel.

Some preliminary treatment is required for most wastes to remove large particulate materials and to allow some suspended solids to settle out. The complete treatment cycle involves both aerobic and anaerobic steps although local requirements will determine which of these is carried out.

Anaerobic waste treatment and biogas production

Many anaerobic digesters consist essentially of closed containers with perhaps provision for gas collection and occasional mixing (Fig. 13.1). Recent innovations include film reactors employing immobilized micro-organisms. Biogas is the product of the anaerobic decomposition of organic material and the most important component produced is methane. Methane is very insoluble, separates readily from a fermentation system and is easily collected and pressurized or liquefied for storage. In addition it is readily combustible and as such is a valuable energy source. The

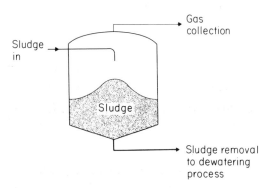

Fig. 13.1. Simplified anaerobic sludge digester.

143

Chapter 13
Waste
treatment
systems and
biodegrad-
ation

composition of biogas varies with the nature and concentration of substrates and the temperature of incubation. The methane content would usually be in the range 60–75% with CO_2 accounting for 25–30% and the remainder being H_2 and N_2. Methane yields of up to 600 litres per kg sludge utilized have been reported with values for dung often lower. In some advanced sewage works and on farms concerned with intensive cattle rearing, methane produced during the anaerobic degradation of wastes may be used as a source of fuel for heating or to propel motor vehicles. The special harvesting of biomass (including water hyacinths from canals and waterways in the USA) for conversion into biogas is practiced in some countries, while in others (such as India and China) small village installations forming biogas from dung and other wastes are a useful addition to the fuel of the local community.

The overall production of methane from organic wastes is a complex fermentation involving a number of groups of micro-organisms which vary in numbers and composition depending upon the effluent composition and other environmental parameters. Growth is slow but based on a 7–14 day throughput, anaerobic digesters reduce the COD typically by about 75%. The initial steps involve the breakdown of polymeric materials (carbohydrates, proteins, fats, etc.). These are carried out by facultative anaerobes as well as anaerobes and form end products such as ethanol and lower fatty acids (acetic, propionic and butyric acids) together with H_2 and CO_2. This process lowers the molecular weight of the wastes and in sewage treatment is often called sludge liquefaction. The methanogenic bacteria (methanogens) are members of the Archaebacteria and are restricted to highly anaerobic environments such as the rumen of cattle and organically enriched sediments. They utilize only a restricted range of substrates, including H_2 and CO_2, acetate, methanol and formate for methane production. Methanogens from the genera *Methanococcus*, *Methanobacterium* and *Methanospirillum* have been isolated from anaerobic treatment plants.

Aerobic waste treatment

Aerobic systems are the most common and range in complexity from the percolating filter (Fig. 13.2), through the various types of activated sludge plants (Fig. 13.3) to the most modern inventions which often use oxygen sparging to produce high rates of microbial activity.

Aerobic digestion requires a large population of actively metabolizing micro-organisms able to degrade both colloidal and soluble organics and with a high rate of conversion to CO_2 and water. The most common aerobic systems are the trickling (per-

144

Chapter 13
Waste
treatment
systems and
biodegrad-
ation

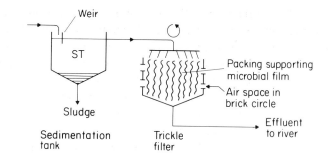

Fig. 13.2. Simple system of effluent treatment by 'trickle filter' ('percolating filter'). The weir at point of effluent input is necessary to maintain still conditions in the sedimentation tank.

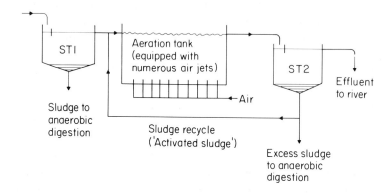

Fig. 13.3. Aerated effluent treatment ('activated sludge' process). ST1 and ST2 are sedimentation tanks.

colating) filters and the activated sludge fermenters. The latter depends upon organisms growing in flocs which are kept in suspension by mechanical agitation or sparging with compressed air. The flocs contain a large number of bacteria including *Zoogloea ramigera* (thought to be responsible for the production of the extracellular polymers which are characteristic of the flocs) and species of *Pseudomonas, Alcaligenes, Achromobacter* and *Brevibacterium.* Ciliated protozoa, such as *Vorticella* spp., are also present and are thought to prey on the bacteria, exerting a control on their numbers and therefore aiding floc stability. Activated sludge plants are the largest application of continuous or semicontinuous fermentation with fresh effluent being added and sludge removed in order to maintain a (roughly) constant population of organisms. The sludge removed is passed to some form of settlement or other dewatering system prior to disposal. This may be carried out in an anaerobic digester or the dewatered sludge is dumped (often at sea), or used on land, potentially as a fertilizer. This latter use is hindered by problems of smell and

145
Chapter 13
Waste
treatment
systems and
biodegrad-
ation

heavy metal content and the potential for the presence of pathogenic micro-organisms. In addition to this problem of sludge disposal the activated sludge process is also expensive in terms of the power requirements for agitation and aeration. The low solubility of oxygen is usually considered to limit the size of the microbial population and hence the overall rate of oxidation.

This low efficiency of oxygen transfer of the traditional activated sludge systems and the large ground area required for their installation has led to developments of new fermentation systems. The ICI deep shaft reactor is based on similar engineering innovations to those used in the 'Pruteen' fermenter but is built into the ground to a depth of about 100 metres. In contrast, a number of companies have built sludge fermenters based on the airlift principle. The Uhde/Hoechst 'Bio-high reactor' is approximately 30 metres high with baffles to improve the mixing characteristics. An alternative strategy involves the use of oxygen rather than compressed air for aeration purposes. A number of commercial plants are in use including the 'Unox' system. This use of oxygen allows smaller aeration and settling tanks to be employed with a saving in capital costs although this has to be offset by the costs of supplying the oxygen and the running of a more complex treatment system.

While the lowering of COD is a prime objective of waste treatment, some effluents also contain high concentrations of nitrogen and phosphorus which if allowed into water courses would encourage eutrophication. The removal of both elements may be achieved by alternating aerobic and anaerobic treatment strategies in the commercial 'Bardenpho', 'A/O' and 'PhoStrip' systems. The degradation of nitrogen containing compounds gives rise to ammonia as a principle product. This ammonia is oxidized to nitrate by chemolithotrophic nitrifying bacteria in an aerobic reactor. This step is slow and rate limiting overall. The next step is anaerobic and relies on the action of denitrifying bacteria which use nitrate and nitrite as electron acceptors and convert these to nitrogen gas which is discharged. The removal of phosphorus involves release into the liquid waste in an anaerobic stage followed by assimilation into bacteria in an aerobic stage. This assimilation appears to involve *Acinetobacter* sp. which accumulate polyphosphate in intracellular granules. The bacterial mass is removed as part of the sludge.

The problems of toxic chemicals

The United States Environmental Protection Agency (USEPA) lists some 60,000 chemicals currently marketed within the USA with a further 1200 or so added to the list on an annual basis. The bulk of these materials probably do not create any hazard to the envi-

146

Chapter 13
Waste
treatment
systems and
biodegrad-
ation

ronment. In the USA alone, however, some 14,000 industrial plants generate some 265×10^6 tonnes of waste materials each year which may be classed as hazardous. Again in the USA there are 4800 on-site treatment or storage facilities for especially toxic materials while the remaining wastes end up on about 93,000 landfill or dump sites. Of these is has been estimated that 16,000 to 20,000 may eventually pose some health risk. In addition, in 1979 some 2×10^6 tonnes of pesticides were used in the Western Hemisphere and in the UK alone the pesticide used was sufficient to produce an average concentration of 30 mg/kg soil over the top 0.5 cm agricultural land. The US Office of Technology Assessment has calculated the bill in the USA for a clean-up operation at $100 billion. There is no evidence to suggest that the problem is any less in other parts of the industrial world.

Inevitably in the handling of this vast amount of toxic waste materials some spills occur. In the USA a study of 950 human populations indicated that some 30% received water from aquifers containing detectable levels of toxic contaminants, especially short chain halogenated hydrocarbons.

Clean up operations may mean removing waste materials, contaminated soils, etc. and subjecting these to incineration (as with the dioxin waste from the Seveso incident). The alternatives include chemical or biological processing. Biodegradation may occur naturally by the action of microbial systems but in many cases the rates of reaction may be too slow to be of practical significance. There is clearly scope for the development of starter cultures by enrichment methods, by conventional genetics, by recombinant DNA technology or by combinations of these strategies to increase the potential of biological solutions to these problems.

Table 13.1 lists some micro-organisms which have been reported to be metabolically able to carry out the biodegradation of some of these compounds. This does not necessarily indicate that these organisms are so involved in any natural environment situation but rather the potential for such an involvement.

The conditions necessary for biodegradation

The key to an assessment of the fate of organic chemicals in the environment is a realistic evaluation of their susceptibility to mineralization (complete conversion to carbon dioxide and water). This is considered later. Photo-oxidation and other abiotic mechanisms are thought to play only a minor role and mineralization is generally achieved only as a result of microbial activity.

Organic compounds may be biodegradable (transformed by biological mechanisms which might lead to complete mineralization), persistent (fail to undergo biodegradation in a particular

Table 13.1. Micro-organisms able to degrade some toxic chemicals.

147

Chapter 13
Waste
treatment
systems and
biodegrad-
ation

Organisms	Toxic chemicals
Pseudomonas spp.	4-alkylbenzoates, alkyl ammonium, alkylamineoxides, anthracene, benzene, hydrocarbons, melathion, naphthalene, methyl naphthalenes, organophosphates, PCBs, p-xylene, p-cymene, parathion, phenanthrene, phenoxyacetates, phenylureas, polycyclic aromatics, rubber, secondary alkylbenzenes, toluene, phenolics, oleaginous materials, pulp byproducts
Alcaligenes spp.	halogenated hydrocarbons, linear alkylbenzene sulphonates, polycyclic aromatics, PCBs
Arthrobacter spp.	benzene, hydrocarbons, pentachlorophenol, phenoxyacetates, polycyclic aromatics
Bacillus spp.	aromatics, long chain alkanes, phenylureas
Corynebacterium spp.	halogenated hydrocarbons, phenoxyacetates
Mycobacterium spp.	aromatics, branched hydrocarbons, benzene, cycloparaffins
Nocardia spp.	hydrocarbons, alkylbenzenes, naphthalene, phenoxyacetates, polycyclic aromatics
Streptomyces spp.	diazinon, phenoxyacetates, halogenated-hydrocarbons
Xanthomonas spp.	hydrocarbons, polycyclic hydrocarbons
Candida tropicalis	PCBs
Cunninghamella elegans	PCBs, polycyclic aromatics
Fusarium solani	propanil

environment or under a specific set of experimental conditions) or recalcitrant (inherently resistant to biodegradation). Most naturally occurring or biogenic compounds are biodegradable while man-made or xenobiotic compounds may be biodegradable, persistent or recalcitrant.

Biodegradation in a particular environment requires the presence of suitable micro-organisms. As discussed later this may involve a complex microbial community. The environment must also be suitable both for the growth of these organisms and for any chemical transformation reaction to proceed at a significant rate. Important factors include the concentration of the toxic chemicals (which most probably will be toxic to the micro-organisms carrying out the transformations), the presence of other substrates and nutrients, temperature, pH, Eh, oxygen concentration, etc. In addition the physical nature of the toxic material is of significance, for example, aerobic transformation of a water insoluble compound will proceed faster in a well mixed and aerated environment (such as surface seawater) than if the same compound is coated on to fine sediment particles in an undersea dumpsite.

The biodegradation of xenobiotics

Because they are man-made and have been developed quite

148

*Chapter 13
Waste
treatment
systems and
biodegrad-
ation*

recently xenobiotic chemicals have been present in the environment for comparatively short periods of time. This in turn means that the microbial communities present in these environments may not have evolved specific mechanisms for their degradation. A number of possible mechanisms exist, however, which may lead to active biodegradation.

The main specificity of enzymes is concerned with their catalytic function, thus a hydrolase does not usually function as an oxidase or a ligase. In contrast many enzymes are much less specific with respect to substrate binding. It is therefore not uncommon for enzymes to bind analogues of their natural substrates which contain xenobiotic functional groups. If these do not greatly alter the charge make up of the active site it is possible for the enzyme to catalyse its particular reaction with the xenobiotic as substrate. The success of this 'gratuitous' metabolism as a biodegradation mechanism depends also on other factors such as the ability of the xenobiotic to act as an inducer and the nature of the product formed (can it be metabolized further, is it toxic, etc.). The metabolism of substituted benzoates illustrates this mechanism. Thus the 3- and 4-trifluoromethylbenzoates were oxidized at a high rate by cultures enriched in the presence of 4-ethyl- and 4-isopropyl-benzoates.

A further mechanism which involves the gratuitous use of existing enzyme systems is cometabolism. A cometabolite does not support the growth of the organism concerned and the products of the transformation are accumulated stoichiometrically. The transformation does require energy consumption of the organism and this is usually measured as an increased uptake of oxygen by the culture. A number of examples of cometabolism are listed in Table 13.2.

The classical approach of the microbiologist and biochemist to the problems of studying the metabolism of a xenobiotic com-

Table 13.2.

Organism	Growth substrate	Cometabolic substrate*	Products
Methylomonas	methane	ethane	ethanol, acetaldehyde, acetic acid
Nocardia	hexadecane	toluene	2,3 dihydroxybenzoic-acid,
	hexadecane	ethylbenzene	phenylacetic acid
Achromobacter	benzoic acid	3-chlorobenzoate	4-chlorocatechol salicylic acid
Corynebacterium	hexadecane	naphthalene	2-hydroxy,
	glucose	anthracene	3-naphthoic acid

*Cometabolism is the transformation of a nongrowth substrate in the obligate presence of a growth substrate or another transformable compound.

pound is to attempt to isolate pure cultures of organisms able to grow at the expense of that particular substrate. If successful then the details of the transformation are elucidated in further experiments often using cell-free preparations. The degradation of a range of compounds, however, has been shown to proceed more readily with a mixed culture of organisms and the utilization of some compounds may not proceed at all in monocultures. A list of examples is given in Table 13.3 and details of the involvement of communities in the degradation of Dalapon and 2-chloropropionamide are given in Figs 13.4 and 13.5. All of these examples hinge upon the use of continuous flow cultures for the enrichment of communities from natural microbial populations.

149
Chapter 13
Waste
treatment
systems and
biodegrad-
ation

Table 13.3. Some compounds degraded by synergistic microbial communities.

dalapon	lontrel
chlorobenzoate	4,4-dichloro-biphenyl
parathion	polychlorinated-biphenyls
isopropyl thiocabamate	cyclohexane
benzoate	linear alkylbenzoate sulphonates
nitrosamines	diazinon
2-(2-methoxy,4-chloro)-phenoxy	propionic acid
alkylphenol ethoxylates	styrene
3,4-dichloropropionanilide	isopropyl phenylcarbamate

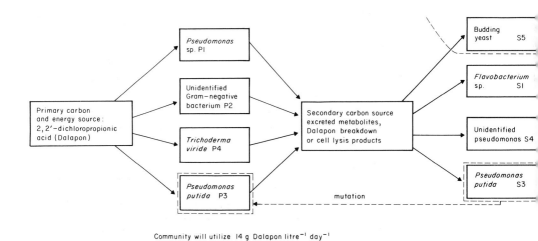

Community will utilize 14 g Dalapon litre^{-1} day^{-1}

Fig. 13.4. Metabolism of Dalapon by a six-membered microbial community.

Plasmids and the evolution of new pathways

While most of the functions essential for the growth of a bacterial cell are encoded in chromosomal genes, the presence of plasmids gives the host cell a survival or growth advantage under some

150

Chapter 13
Waste
treatment
systems and
biodegrad-
ation

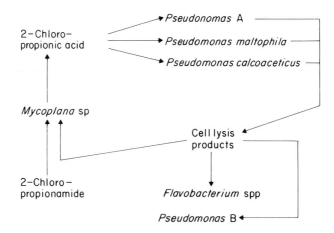

Fig. 13.5. 2-Chloropropionamide utilization.

conditions. Genes coding for some enzymes essential for the biodegradation of a number of organic compounds are plasmid borne (Table 13.4) and organisms have been constructed to degrade difficult wastes. For example a patented process has been developed with a strain of *Pseudomonas putida* constructed to contain plasmids coding for the breakdown of octane, xylene, metaxylene and camphor. This organism is claimed to be useful in the clean up of oily water discharges and oil spills.

Table 13.4. Some plasmids concerned in biodegradation.

Chemical species	Plasmid	Chemicals transformed
alkanes	OCT	octane, decane
aromatic and polyaromatic	TOL (pWWO)	xylenes, toluene
hydrocarbons and	NAH	naphthalene
metabolic products	SAL	salicylate, benzoate
terpenes	CAM	camphor
alkaloids	NIC	nicotine, nicotinate
chlorinated hydrocarbons	pJP1	2,4 D
	pAC21	p-chlorobiphenyl 3-chlorobenzoate

The possible role of plasmids in the natural selection of organisms able to degrade xenobiotics is illustrated in the example of the utilization of chlorobenzoates by *Pseudomonas* spp. (Fig. 13.6). The combined metabolic activities of strains B13 and mt2 allowed the degradation of a number of chlorobenzoates. Thus strain B13 contained a very selective benzoate 1,2 dioxyganase which would not catalyse the hydroxylation of halogenated benzoates together with a catechol 1,2 dioxygenase which could

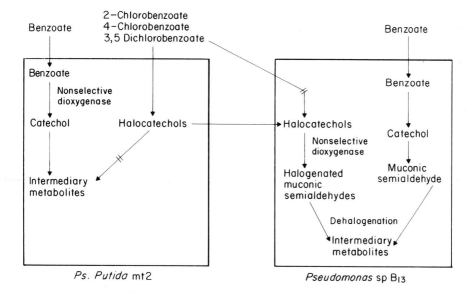

Fig. 13.6. Utilization of Chlorobenzoates by *Pseudomonas* spp.
Note: The transfer of nonselective Benzoate 1,2 Diaxygenase from mt2 to B$_{13}$ on plasmid pWWO is responsible for production of a single organism able to utilize these chlorobenzoates.

transform 4-chlorocatechol. In contrast, strain mt2 contained a benzoate dioxygenase of broader specificity but nevertheless also would not grow on the halogenated substrates due to the stringent specificity of the catechol dioxygenase. In continuous cultures containing both organisms good growth and substrate oxidation was achieved. Such mixed cultures also lead to the evolution of a single organism possessing the required properties from both parents. This was achieved by the transfer of the gene coding for the nonselective benzoate dioxygenase from mt2 to B13 on the TOL (pWWO) plasmid.

Using a similar experimental strategy, organisms have been selected in continuous flow cultures which are able to degrade 2,4,5T. Thus waste dumpsite material was enriched in the presence of bacterial strains carrying plasmids coding for the utilization of camphor, toluene, salicylic acid, p-chlorobiphenyl, 3-chlorobenzoate and 4-chlorobenzoate and fed a mixture of substrates including toluate, salicylate, chlorobenzoates and low concentrations of 2,4,5T. Over an incubation period of some 10 months the concentration of 2,4,5T was steadily increased. Strains were evolved able to degrade up to 1.5 g/l 2,4,5T.

There is also a clear potential for the use of recombinant DNA technology to construct strains with special degradative capacities. Research in this area is at an early stage but has a high potential for the future.

152

Chapter 13
Waste
treatment
systems and
biodegrad-
ation

How to evaluate biodegradability

A knowledge of the degree to which any new compound or formulation of compounds is biodegradable is required as part of an assessment of the likely impact on the environment if released. Both the OECD (Organization for Economic Cooperation and Development and the USEPA have advocated a multitiered approach to biodegradation testing. Thus a number of requirements may have to be met in order to make conditions suitable for biodegradation and these may be time consuming and expensive to achieve. In contrast simple and inexpensive tests may be adequate for the breakdown of readily biodegradable compounds. Multitiered tests therefore start with simple procedures and move to more complex experimentation only if the need arises.

Tier 1 consists of simple screening tests with the aim of eliminating those compounds unlikely to cause problems due to persistence. In the OECD guidelines a compound is considered to be biodegradable if it is completely mineralized within 28 days when used as sole carbon source by a small inoculum of sewage organisms without prior exposure of the inoculum to the substrate. Evidence for mineralization may be oxygen consumption, carbon dioxide evolution, removal of dissolved organic carbon, etc. These represent very stringent conditions of testing. Even if a compound passes this test it will require to be tested further for toxicity before approved for use.

The failure of an organism in Tier 1 may be due solely to the stringent conditions imposed, alternatively the compound may be persistent or recalcitrant. The Tier 2 procedures consist of enrichment and selection systems. The methods involved range from simple shake flasks and fill and draw activated sludge systems to chemostat enrichment cultures and continuous flow activated sludge fermenters. For example a chemostat culture might be operated at a low specific growth rate with a multicomponent carbon source while receiving regular new inocula from a variety of different sources, the compound under test being used at low concentration to begin with and then this concentration being increased gradually. Since a multicomponent medium would usually be applied then specific chemical analyses for the compound under test or the availability of a radiolabelled substrate would be required to assess its breakdown. It has been suggested that a fermentation system should be run for a period of three months or more unsuccessfully before recalcitrance may be assumed.

Tier 3 involves an assessment of the degree of biodegradation. If a successful culture has been obtained in a Tier 2 enrichment, then this may be used as an inoculum for a Tier 1 mineralization test. Alternatively a radiolabelled substrate might be employed.

The most complete results would probably be obtained by determining the carbon balance in a steady state chemostat culture.

Tier 4 assesses the kinetics of biodegradation i.e. the predicted rates of mineralization in a given environment and may involve radiolabelled substrates and sophisticated analytical methods.

153

Chapter 13

Waste

treatment

systems and

biodegrad-

ation

Chapter 14. Regulation and safety

All industrialized countries have regulations governing safety at work, pharmaceuticals, food and drink, environmental protection and pollution control which are relevant to the biotechnological industries. Included are guidelines concerning the use of pathogens and of organisms containing recombinant DNA sequences and their use and containment.

Process design and containment

Fermentation processes vary in their requirements. In general the old and established processes including wastewater treatment and some food processes require only a minimum level of sterility and containment. For example in the treatment of wastewater a natural population of organisms is enriched and employed often in open vessels and without harm to the operators or to the surrounding environment. The production of beer and the milk fermentations require a clean and hygienic environment and a rapid fermentation using healthy starter cultures to ensure the integrity of the fermentations. Both types of process will tolerate low levels of contaminating organisms. The production of both bulk and speciality chemicals such as ethanol, citric acid, xanthan gum and amino acids are also low on the scale of containment requirements with the need in some instances for sterile processing being dictated by the prevailing food regulations and a need to maintain the integrity of the fermentation system. Single cell protein production also requires only low levels of containment, since organisms authorized for animal or human consumption are unlikely to be pathogenic, but requirements for sterility may be high. This is especially true if continuous fermentations operating for long periods of time are to be employed. Antibiotic fermentations are subject to a higher level of control with sterile working practices required for the production of pharmaceutical products but generally without the need for containment of the fermentation and processing systems. At the highest level of sophistication are fermentations employing pathogens or organisms with recombinant DNA inserts which require not only sterile practices to protect the product but also containment to protect the process workers and the environment from accidental leaks. These different categories of processes clearly overlap and a general scheme is shown in Fig. 14.1.

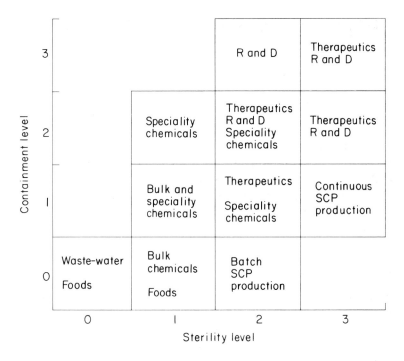

Fig. 14.1. Containment and sterility requirements of some biotechnological processes.

The degree of containment required in any instance will usually be determined by regulatory authorities. In the UK the Advisory Committee on Dangerous Pathogens (ACDP) has drawn up guidelines for the use of pathogenic organisms while the use of organisms with recombinant DNA inserts, initially governed by the Genetic Manipulation Advisory Group (GMAG), is now the responsibility of the Health and Safety Executive. Many industrialized countries have followed the lead given by the National Institutes of Health in the USA which through the Recombinant DNA Advisory Committee (RAC) produced detailed guidelines concerning the handling of genetically engineered organisms from laboratory experiments to largescale industrial use. Initially these guidelines were extremely strict but they have been relaxed as practical experience has indicated that the fears of safety problems associated with genetically engineered organisms were overstated. Nevertheless this is kept under review and there are a number of pressure groups and activists whose presence will guarantee that it will remain subject to public awareness and scrutiny.

An indication of containment requirements is given in Table 14.1 which groups together the guidelines and codes of the UK and USA within some arbitrary categories. Category 0 demands the implementation of Good Manufacturing Practice (GMP) and

Table 14.1.

Risk		Existing codes		
category	Examples	GMAG	ACDP	NIH (USA)
3 (high)	Dangerous pathogens, some genetically-engineered organisms	IV	4	P3
2 (medium)	Experimental and R&D	III/II	3/2	P2/P1
1 (low)	'Proven processes'	I	1	—
0	Waste treatment food	—	—	—

otherwise is similar to process plant in the chemical industries. For category 1 'proven processes' the likely standards also include in place cleaning of vessels and equipment, protection against aerosols by careful choice of equipment, the absence of handling of live organisms and process area containment in addition to the general requirements of GMP. Category 2 with moderate containment requires all the standards of category 1 and in addition a higher quality of fermentation system with secondary seals, the incineration or filtration of all extract air from the fermentation plant and provision for the collection and sterilization of condensate. In addition the process building would be maintained under a negative pressure and access would be restricted to essential personnel only. The highest category 3 demands all those for 2 and in addition separate containment around all process equipment with personnel access only in protective clothing or after sterilization of the working environment. In addition to an overall negative pressure to prevent leaks outwards, the process building itself would also be contained with all exhaust air being filtered or incinerated. The liquid effluent from these high category buildings would also be carefully monitored with additional sump tanks and sterilization facilities if required. While category 3 systems are necessary for research they are likely to prove too expensive for production purposes except for small volumes of very high value products.

Food and drug safety

All new products designed for food use or health care for the human, food animal or domestic animal populations are required to be tested for safety before marketing is approved. A new drug destined for human use is first tested extensively in animals in order to determine toxicity, teratogenicity, efficacy, etc. in long and expensive experimentation. Only at this stage are limited trials on human volunteers authorized, leading finally to full scale clinical trials and then marketing. The costs of bringing a new drug through all of these stages has risen steadily in recent years to over $100 million, while the time taken has increased from about one year in 1960 to over seven years now. These costs have to be covered by sales, often during a limited period of patent protection, following marketing approval. A new food product must undergo a similar programme of testing. For example approval for the sweetener 'Aspartame' took over ten years while research and approval for the *Fusarium* 'Mycoprotein' cost £30 million.

These regulations apply equally to products made with and without the use of recombinant DNA. Most of the new products of recombinant DNA technology likely to be marketed in the near future are proteins. Human insulin, human somatotrophin (growth hormone) and interferon-alpha, are current examples with interleukin-2, alpha-antitrypsin, tissue plasminogen activator and a number of viral antigens for the near future. Existing legislation was developed largely for small molecular weight pharmaceuticals and is not easily extended to these new products. In the absence therefore of detailed regulations, each new protein is considered on a case by case basis. The final specification of any product depends on factors such as the route of administration (e.g. orally or by injection) and the duration of therapy likely to be prescribed. The type of specification required will then include purity ($> 95\%$), absence of microheterogeneity and the presence of a consistent three-dimensional structure, absence of the chemicals used in purification, absence of endotoxin (of special significance if *E. coli* is used as the production organism), absence of bacterial and plasmid DNA and the absence of immunoglobulins (if these have been used in an affinity purification scheme).

Environmental release

The deliberate release of micro-organisms into the environment is not a new innovation. For example and as indicated in an earlier chapter, insecticide preparations based on the *Bacillus thuringiensis* endotoxin are in widespread application. A recombinant *Pseudomonas fluorescens* containing the *B. thuringiensis* gene coding for the endotoxin, however, is restricted to laboratory and glasshouse tests pending approval on a wider scale. The same applies to a number of other organisms modified by recombinant DNA methods, including a *Pseudomonas syringae,* the so-called 'ice minus' bacterium, which has been the subject of lengthy litigation in the USA. In the absence of approved guidelines all new applications are dealt with on a case by case basis. Guidelines for environmental release are being formulated, however, and in the USA this is the responsibility of the Biotechnology Science Coordinating Committee. The policies likely to be adopted will include a review of the organisms prior to their release into the environment. For example, all organisms containing DNA from dissimilar source organisms (so-called intergeneric combinations) will be subject to review by the Environmental Protection Agency before approval for small scale (< 10 acres) or larger scale use. A similar review machinery will apply to those organisms formed by recombinant methods which do not involve intergeneric combinations where the source organisms are known pathogens. Following any release it is essential that organisms containing recombinant DNA (and the DNA itself) should be monitored for their spread in the environment. This requires a knowledge of the survival potential of these organisms and the ability to detect these and/or the recombinant DNA. Sensitive and specific detection systems involving methods such as gene probes and monoclonal antibody-linked probes are likely to be employed. Similar problems apply to plants modified by recombinant methods, some of which already are the subject of largescale glasshouse trials. For example over 5000 tobacco plants made resistant to the herbicide atrazine by insertion of the detoxifying enzyme glutathione-S-transferase have been 'field tested' in such restrictive conditions.

Further reading

Chapter 1

Hacking A.J. (1986) *Economic Aspects of Biotechnology.* Cambridge University Press, Cambridge.

Chapter 2

Enei H. & Hirose Y. (1984) Recent research on the development of microbial strains for amino acid production. *Biotechnology & Genetic Engineering Reviews*, **2**, 101–120.

Feinberg E.L., Ramage P.I.N. & Trudgill P.W. (1980) The degradation of n-alkyl-cycloalkanes by a mixed bacterial culture. *Journal of General Microbiology*, **121**, 502–511.

Hager P.W. & Rabinowitz J.C. (1986) Translational specificity in *Bacillus subtilis*. In *Molecular Biology of the Bacilli*, Vol 2, pp. 1–32 (Ed. D. Dubnau). Academic Press, Orlando, Florida.

Randall L.L. & Hardy S.J.S. (1984) Export of protein in bacteria. *Microbiological Reviews*, **48**, 290–298.

Schleif R. (1986) *Genetics & Molecular Biology.* Addison Wesley, Reading, Massachusetts.

Slater J.H. & Bull A.T. (1982) Environmental microbiology: biodegradation. *Philosophical Transactions of the Royal Society of London*, **B297**, 575–579.

Williams S.T., Goodfellow M. & Vickers J.C. (1984) New microbes from old habitats? *Symposium of the Society for General Microbiology*, **36**, 219–256.

Yamane K. & Maruo B. (1980) *B. subtilis*—amylase genes. In *Molecular Breeding and Genetics of Applied Micro-organisms*, pp. 117–123 (Ed. K. Sakaguchi & M. Ouishi). Academic Press, London & New York.

Chapter 3

Stierley S.L. & Young T.W. (1986) Genetic manipulation of commercial yeast strains. *Genetic Engineering & Biotechnology Reviews*, **4**, 1–38.

Doi R.H. (1984) Genetic Engineering in *Bacillus subtilis*. *Biotechnology & Genetic Engineering Reviews*, **2**, 121–156.

Harris T.J.R. (1983) Expression of eukaryotic genes in *E. coli*. In *Genetic Engineering*, **4**, pp. 128–185 (Ed. R. Williamson). Academic Press, London.

Hopwood D.A., Bibb M.S., Bruton C.J., Chater K.F., Feitelson J.S. & Gil J.A. (1983) Cloning *Streptomyces* for antibiotic production. *Trends in Biotechnology*, **1**, 42–48.

Fayerman J.T. (1986) New developments in gene cloning in antibiotic-producing micro-organisms. *Biotechnology*, **4**, 786–789.

Glover D.M. (1984) *Gene Cloning, the Mechanics of DNA Manipulation.* Chapman & Hall, London & New York.

Kellog S.T., Chakrabarty D.K. & Chakrabarty A.M. (1981) Plasmid assisted molecular breeding, a new technique for enhanced biodegradation of persistent toxic chemicals. *Science*, **214**, 1133–1135.

Old R.W. & Primrose S.B. (1985) *Principles of Gene Manipulation: an Introduction to Genetic Engineering*, 3rd Edition. Blackwell Scientific Publications, Oxford.

Outtrup H. & Norman B.E. (1984) Properties and application of a thermostable maltogenic amylase produced by a strain of *Bacillus* modified by recombinant DNA techniques. *Die Starke*, **36**, 405–411.

Schoner R.G., Ellis L.F. & Schoner B.E. (1985) Isolation and purification of protein granules from *Escherichia coli* cells overproducing bovine growth hormone. *Biotechnology,* 3, 151–154.

Valenmala P., Medina A., Rutter W.J., Ammerer G. & Hall B.D. (1982) Synthesis and assembly of hepatitis B virus antigen peptides in yeast. *Nature,* **298,** 347–350.

Chapter 4

Glacker M.W., Fleischaker R.J. & Sinskey A.J. (1983) Mammalian cell culture: engineering principles and scale-up. *Trends in Biotechnology,* 1, 102–108.

Hubbard R. (1983) Monoclonal antibodies: production, properties and applications. In *Topics in Enzyme and Fermentation Technology,* Vol 7, pp. 196–263 (Ed. A. Wiseman). Ellis Horwood, Chichester.

Kucherlapati R.S. (1984) Induction of purified genes into animal cells. *ASM News,* **50,** 49–53.

Shih C. & Winberg R.A. (1982) Isolation of a transforming sequence from a human bladder carcinoma cell line. *Cell,* **26,** 67–78.

Smith G.L., Mackett M. & Moss B.M. (1984) Recombinant vaccinia viruses as new vaccines. *Biotechnology & Genetic Engineering Reviews,* 2, 383–407.

Wilson T. (1984) More protein from mammalian cells. *Biotechnology,* 2, 753–755.

Chapter 5

Barton K.A. & Brill W.J. (1983) Prospects in plant genetic engineering. *Science,* **219,** 671–675.

Fowler M.W. (1984) Plant cell culture: natural products and industrial applications. *Biotechnology and Genetic Engineering Reviews,* 2, 41–48.

Hussey G. (1983) In vitro propagation of horticultural and agricultural crops. In *Plant Biotechnology* (Ed. S.H. Mantell & H. Smith), pp. 111–138. Cambridge University Press, Cambridge.

Lindsey K. & Yeoman M.M. (1983) Novel experimental systems for studying the production of secondary metabolites by plant cell culture. In *Plant Biotechnology* (Ed. S.H. Mantell & H. Smith), pp. 39–66. Cambridge University Press, Cambridge.

Mantell S.H., Mathews J.A. & McKee R.A. (1985) *Principles of Plant Biotechnology, An Introduction to Gentic Engineering in Plants.* Blackwell Scientific Publications, Oxford.

Chapter 6

Atkinson B. & Mavituna F. (1983) *Biochemical Engineering and Biotechnology Handbook.* Macmillan, London.

Bailey J.E. & Ollis D.F. (1977) *Biochemical Engineering Fundamentals.* McGraw Hill, New York.

Crueger W. & Crueger A. (1984) *Biotechnology: A Textbook of Industrial Microbiology.* Sinauer Assoc., Sunderland, Massachusetts.

Hatch R.T. (1975) Fermenter design. In *Single Cell Protein* (Eds S.R. Tannenbaum & D.I.C. Wang), Vol 2, pp. 46–68. MIT Press, Cambridge, Massachusetts.

Laskin A.J. (1977) Single cell protein. In *Annual Reports on Fermentation Processes,* Vol 1, pp. 151–175. Academic Press, New York.

Stanbury P.F. & Whitaker A. (1984) *Principles of Fermentation Technology.* Pergamon Press, Oxford.

Wang D.I.C., Cooney C.L., Demain A.L., Dunill P., Humphrey A.E. & Lilly M.D. (1979) *Fermentation and Enzyme Technology.* John Wiley & Sons, New York.

Chapter 7

Atkinson B. & Mavituna F. (1983) *Biochemical Engineering & Biotechnology Handbook.* Macmillan, London.

Cooper A.R. (Ed.) (1983) *Ultrafiltration Membranes and Applications.* Plenum Press, New York.

Goldblith S.A., Rey L. & Rothmayr W.W. (Eds) (1984) *Freezedrying and Advanced Food Technology.* Academic Press, New York.

Scouten W.H. (Ed.) (1985) *Affinity Chromatography: bioselective adsorption on inert matrices.* John Wiley, New York.

Stanbury P.F. & Whittaker A. (1984) *Principles of Fermentation Technology.* Pergamon Press, Oxford.

Chapter 8

Broughty-Boye G.C., Maman J., Marian J-C. & Chuay P. (1984) Biosynthesis of human tissue plasminogen activator by normal cells. *Biotechnology* , **2**, 1058–1062.

Cheetham P.J.S. (1985) The application of enzymes in industry. In *Handbook of Enzyme Technology*, 2nd Edn., pp. 274–379 (Ed. A. Wiseman). Ellis Horwood, Chichester.

Denehy J.P. & Wolnak B. Eds (1980) *Enzymes, the Interface Between Technology and Economics.* Marcel Dekker, New York.

Powell L.W. (1984) Developments in immobilised-enzyme technology. *Biotechnology & Genetic Engineering Reviews*, **2**, 409–438.

Priest F.G. (1984) *Extracellular Enzymes.* Van Nostrand Reinhold, Wokingham, UK.

Sundaram P.V. (1982) Analytical applications for routine use with immobilized enzyme nylon tube reactors. *Microbial and Enzyme Technology*, **4**, 290–298.

Ueda S., Saha B.C. & Koba Y. (1984) Direct hydrolysis of raw starch, *Microbiological Sciences*, **1**, 21–24.

Chapter 9

Crueger W. & Crueger A. (1983) *Biotechnology: A Textbook of Industrial Microbiology.* Sinauer Assoc., Sunderland, Massachusetts.

Enei H., Shibai H. & Hirose Y. (1982) Amino Acids and nucleic-acid related compounds. *Annual Reports on Fermentation Processes*, **5**, 79–100.

Hough J.S. (1984) *The Biotechnology of Malting & Brewing.* Cambridge University Press, Cambridge.

Krumphanzi V., Sikyta B. & Vanek Z. (Eds) (1982) *Overproduction of Microbial Products.* Academic Press, London.

Chapter 10

Rothman H., Greenshields R.N. & Calle F.R. (1983) *The Alcohol Economy: Fuel Ethanol and the Brazilian Experience.* Frances Pinter, London.

Goldstein I.S. (1981) *Organic Chemicals from Biomass.* CRC Press, Florida.

Hacking A.J. (1986) *Economic Aspects of Biotechnology.* Cambridge University Press, Cambridge.

Chapter 11

Demain A.L. & Solomon N.A. (1983) *Antibiotics: Handbook of Experimental Pharmacology.* Springer Verlag, Berlin.

Rosazza J.P. (1982) *Microbial Transformations of Bioactive Compounds.* CRC Press, Florida.

Perlman D. (1979) Microbial production of antibiotics. In *Microbial Technology* (Eds H.J. Peppler & D. Perlman), **Vol. 1**, pp. 241–280. Academic Press, New York.

Chapter 12

Alfermann A.W., Bergman W., Figur C., Helmbold U., Schwantag D., Sculler I. & Reinhard E. (1983) Biotransformation of m-methoxydigitoxin and β-ethyldigoxin by cell cultures of *Digitalis lanata*. In *Plant Biotechnology* (Ed. S.H. Mantell & H. Smith), pp. 67–74. Cambridge University Press, Cambridge.

Aronson A.L., Beckman W. & Dunn P. (1986) *Bacillus thuringiensis* and related insect pathogens. *Microbiological Reviews*, **50**, 1–24.

Ammirato P.V., Evans D.A., Flick C.E., Whitaker R.J. & Sharp W.R. (1984) Biotechnology and agricultural improvements. *Trends in Biotechnology*, **2**, 53–58.

Cheung A.K. & Kupper M. (1984) Biotechnological approach to a new foot-and-mouth disease virus vaccine. *Biotechnology & Genetic Engineering Reviews*, **1**, 223–260.

Deacon J.W. (1983) *Microbial Control of Plant Pests and Diseases.* Van Nostrand Reinhold, Wokingham, UK.

Fowler M.W. (1983) Commercial applications and economic prospects of mass plant cell culture. In *Plant Biotechnology*, pp. 3–38 (Ed. S.H. Mantell & H. Smith), Cambridge University Press, Cambridge.

Hammer R.E., Pursel V.G., Rexsroad Jr C.E., Wall J.R., Bolt D.J., Ebert K.M., Palmiter R.D. & Brinster R.L. (1985) Production of transgenic rabbits, sheep and pigs by microinjection. *Nature*, **315**, 680–683.

Marston F.A.O., Lowe P.A., Doel M.T., Schoemaker J.M., White S. & Angel S. (1984) Purification of calf prochymosin (prorennin) synthesized in *Escherichia coli. Biotechnology*, **2**, 162–166.

Messing J. (1983) The manipulation of zein genes to improve the nutritional value of corn. *Trends in Biotechnology*, **1**, 55–58.

Chapter 13

Crueger W. & Crueger A. (1984) *Biotechnology: A Textbook of Industrial Microbiology.* Sinauer Assoc., Sunderland, Massachusetts.

Jain R.K. & Sayler G.S. (1987) Problems and potential for *in situ* treatment of environmental pollutants by engineered micro-organisms. *Microbiological Sciences*, **4**, 59–63.

Kilbane J.J., Chaterjee D.K. & Chakrabarty A.M. (1983) Detoxification of 2,4,5-trichlorophenoxyacetic acid from contaminated soil by Pseudomonas cepacia. *Applied & Environmental Microbiology*, **45**, 1697–1700.

McCormick D. (1985) 'One bug's meat . . .'. *Biotechnology*, **3**, 429–435.

Magruder G.C. & Gaddy J.L. (1981) Production of farm energy from biomass. In *Advances in Biotechnology*, **2**, 269–274.

Slater J.H. & Bull A.T. (1982) Environmental Biotechnology: biodegradation. *Philosophical Transactions of the Royal Society*, **B297**, 575–597.

Stanbury P.F. & Whittaker A. (1983) *Principles of Fermentation Technology.* Pergamon Press, Oxford.

Chapter 14

Brown C.M. & Duffus J.H. (1984) Health impact of biotechnology. *Swiss Biotechnology*, **2**, 7–32.

Stotzky G. & Babich H. (1986) Survival of, and genetic transfer by, genetically engineered bacteria in natural environments. *Advances in Applied Microbiology*, **31**, 93–138.

Levin M.A., Seidler P., Borquin A.W., Fowler J.R. & Barry T. (1987) EPA developing methods to assess environmental release. *Biotechnology*, **5**, 38–45.

Index